Adapted from a map produced a[...]
summarizing environmental proble[...]
tions Conference on the Human E[...]

Civic Garden Centre
Library

Area of overfishing		Offshore oil wells	Projected supersonic air lane
Thermal pollution from nuclear power plant			Land largely impoverished by misuse
Dumping site for radioactive waste		Urban air pollution	Strip-mining
Polluted water		Industrial complex or power plant	Deforested area
Mercury pollution		Overuse of pesticides or herbicides	Road
Major dam (built or projected)		Offshore waste disposal	Pipeline (new or projected)
			Coral reef destruction

Earth,
the Great
Recycler

Also by Helen Ross Russell

City Critters
Clarion the Killdeer
Small Worlds: A Field Trip Guide
Soil: A Field Trip Guide
Ten-Minute Field Trips
The True Book of Buds
The True Book of Springtime Seeds
Water: A Field Trip Guide
Winter: A Field Trip Guide
Winter Search Party: A Guide to Insects and Other Invertebrates

Earth, the Great Recycler

by

HELEN ROSS RUSSELL

*illustrated with photographs
and with line drawings
by Gene Garone*

Publishers since 1798

THOMAS NELSON INC.
NASHVILLE / NEW YORK

Photographs not otherwise credited are by the author.

Text copyright © 1973 by Helen Ross Russell

Line drawings copyright © 1973 by Gene Garone

All rights reserved under International and Pan-American Conventions. Published in Nashville, Tennessee, by Thomas Nelson Inc. and simultaneously in Don Mills, Ontario, by Thomas Nelson & Sons (Canada) Limited. Manufactured in the United States of America.

Second Printing, April 1974

Library of Congress Cataloging in Publication Data

Russell, Helen Ross, 1915–
  Earth, the great recycler.

  SUMMARY: Describes the physical, chemical, and biological science concepts involved in cycling and recycling.
  Includes bibliographical references.
  1. Ecology. 2. Earth sciences. 3. Biogeochemistry. 4. Cycles. [1. Earth sciences. 2. Ecology. 3. Recycling (Waste)] I. Garone, Gene, illus. II. Title.
QH541.13.R87   574.5   73–7823
ISBN 0–8407–6268–2

In memory of E. Laurence
Palmer, who was equally at home
in the farms and woods of upstate
New York, on the ranches of New
Zealand, in the mountains of Mexico, in Europe and Africa, and in
any other part of Earth's ecosystem because of his sensitivity to
those universal interrelationships
that constitute Earth's recycling
mechanisms.

Contents

Earth,
the Great
Recycler

CHAPTER ONE

Earth, Our Home

Earth, one of the nine planets in the solar system, is our home. It is also the home of hundreds of thousands of kinds of plants and animals. It is a sphere 7,922 miles in diameter which travels around the sun at a speed of approximately 66,600 miles per hour and rotates on its axis once every twenty-four hours.

When astronauts looked back at Earth from outer space, they rejoiced at its beauty. They took photographs that showed the spinning globe with its blue oceans and white clouds. When they arrived at the moon, the lunar scene was entirely different. Dry rocks, lunar dust,

Hydrosphere, lithosphere, and atmosphere combined support the biosphere. Here at Boone Lake in East Tennessee cultivated land adjoins wild land, forming an area well adapted to both wildlife and crop production in a spot of great beauty.

Courtesy Tennessee Valley Authority

A view of the lunarscape showing scientist-astronaut Harrison H. Schmitt working at the Lunar Roving Vehicle on the Apollo 17 trip. Without atmosphere, the sky looks black. Without water or air, there is nothing to carve and reshape the surface.

Courtesy NASA. Photo taken by astronaut Eugene A. Cernan.

If it weren't for the low pull of gravity, astronauts would have great difficulty moving with all the gear necessary for survival on the moon. They certainly couldn't carry two packages of equipment, as Edwin E. Aldrin, Jr., is doing in this picture taken during the Apollo 11 mission.

Courtesy NASA. Photo taken by astronaut Neil A. Armstrong.

craggy peaks, deep craters dominated the landscape. There were no cooling waters. There was no atmosphere. There were no plants. There was no life.

When the astronauts started on their homeward journey, Earth looked even better. Although their spaceship was equipped with some recycling equipment, most of the life-giving substances that were packed in it for their use would be converted to other substances if they stayed in outer space too long. Things that are often taken for granted on Earth became tremendously important in outer space.

We breathe 20,000 to 25,000 times every day, removing oxygen from the air each time we inhale. In our bodies the oxygen combines with digested food, and in the process water and carbon dioxide are formed, which we discharge when we exhale. Green plants take the water and carbon dioxide and use them to make food, releasing oxygen in the process. This exchange is part of the Earth's recycling system.

Oxygen is one of the Earth's building blocks. We call these building blocks elements. Elements are substances that cannot be broken into anything else by physical or chemical means. They can only be changed by atomic fission or fusion. The smallest unit of an element is called an atom.

Elements are also the building blocks of the moon and of the other planets. There are ninety-two naturally occurring elements on Earth. They combine with each other to form compounds, such as water and carbon dioxide.

The material the Earth is composed of is called matter. All matter is either an element or a compound.

Earth is different from all the other members of the solar system in two ways. First, its size and mass give it a pull of gravity that is strong enough to hold on to gases and liquids. Second, its distance from the sun and its twenty-four-hour day keep it from being scorching hot like the planet Mercury or freezing cold like the planet Pluto. The surface of the moon, for instance, is scorching hot during its 236-hour day and freezing cold during its 236-hour night.

The sun is the source of energy for Earth and all the rest of the solar system. It is extremely hot, with a surface temperature of 6,000° C. (about 10,800° F.) and an interior temperature estimated at 20,000,000° C. (about 36,000,000° F.) This heat comes from atomic energy created as atoms of hydrogen fuse to form helium.

At temperatures this high the molecules of all substances move rapidly, and all the matter turns into gases. A substance is a gas, a liquid, or a solid because of the rate at which its molecules are moving. The faster the molecules move, the hotter the substance is.

It is hard to think of materials like iron and gold as gases, yet it is possible to turn either of them into a liquid here on Earth by heating them enough. Still more heat would vaporize them. Mercury is a metal that exists as a liquid at Earth's temperatures. When it is heated to 357° C. (675° F.), it turns into a gas which is used in mercury vapor lights. It must be chilled to −38° C. (−36° F.) to become a solid.

Each substance has its own boiling point, the temperature at which it changes from a solid to a gas, and its own melting point, the temperature at which it changes from a solid to a liquid. At the temperatures that exist naturally on Earth, a few substances are gases; some others are liquids; and many are solids.

Most of the gases form a layer about four hundred

miles thick, which surrounds the earth. This mixture of gases is called the air or the atmosphere. The word *atmosphere* comes from two Greek words which mean "a ball of vapor." Most of the gases of the atmosphere are concentrated next to the earth in the lower ten to thirty miles.

The solid material of Earth exists as rocks. They, too, form a continuous sphere, and this sphere is called the lithosphere (*lithos* is a Greek word for "stone").

The rocks of Earth's surface may be broken into small pieces forming clay, sand, or soil, or they may be massive. They may form mountain ranges or plains or dip down to form deep valleys. They may be covered with water, ice, or plants.

Most of the liquid of Earth is water. Water also forms a continuous sphere, called the hydrosphere. More than 70 percent of Earth's surface is covered by oceans, lakes, ponds, rivers, or brooks. In addition, water can always be found under the areas of dry land. It may be only a few inches below the surface, as it is near lakes and oceans, or it may be thousands of feet beneath a desert.

The water in the lithosphere is called ground water, and the level at which it is found is the water table. Regardless of how deep the water table sinks, Earth's waters form a continuous sphere.

The terms *atmosphere, lithosphere,* and *hydrosphere* are intended to describe the wholeness with which gaseous, solid, and liquid substances circle the globe. However, they are confusing, for they suggest that each represents a different layer, and that each layer is always in the same position. This is not the case. All three spheres are interwoven.

Air fills all the holes in the soil and the cracks and crannies in rocks, and it dissolves in water, so it penetrates both the lithosphere and the hydrosphere.

The terms *weather* and *climate* apply to conditions of the atmosphere, but both weather and climate are influenced in large part by the hydrosphere and lithosphere. Air moving over lakes and oceans picks up water vapor; air moving up high mountains is cooled; air moving over rocks is heated differently from air moving over water.

Water vapor is a part of the atmosphere, but it condenses to form liquid water, which floats through the atmosphere as fog or clouds, then falls as rain and sinks into the lithosphere.

The hydrosphere in the form of running water and the atmosphere in the form of wind both shape the lithosphere by breaking, cutting, grinding, sorting, carrying, and depositing solid materials. The moon has no hydrosphere or atmosphere to wear down and move parts of its lithosphere, which is partly why the lunar landscape is different from Earth's.

The form and composition of Earth's lithosphere determines whether the water of the hydrosphere will be above it, below it, or within it. Dust, sand, soil, and other rock particles from the lithosphere are carried in both the hydrosphere and the atmosphere.

The interwoven, interacting forms of matter that make up Earth influence each other, combine with each other, and constantly change and cause change. They provide the environment for living things.

Sometimes the living things are called the biosphere, which means "sphere of life." The biosphere is interwoven throughout the lithosphere, hydrosphere, and atmosphere. It is made up of many different forms, ranging in size from plants and animals that are barely visible with the strongest microscopes to animals like whales and elephants and plants like *Sequoia* trees and giant seaweeds.

Life is a unique state. We know about it because we are alive and experience aliveness. Like all living things, we are born, respond to the environment, use food, move, grow, repair injuries, reproduce, relate to other organisms, and die. These are things that rocks and water and other nonliving substances cannot do. These are things we observe but do not fully understand.

In spite of the differences between living and nonliving things, the biosphere is tied into the lithosphere, hydrosphere, and atmosphere. The bodies of all living things are composed of elements and compounds in the form of solids, liquids, and gases. All life would quickly disappear if the atmosphere or hydrosphere of the world ceased to exist. Many living things are dependent on the lithosphere.

Just as the biosphere is influenced by the environment, so it influences the environment. Plants and animals use the materials from the environment for their bodies; they help shape the land; they change the atmosphere; and they use the water. All living and nonliving things are interrelated.

Earth is special because it can support life. Earth is special because it has life.

CHAPTER TWO

The Building Blocks

Stop and mentally freeze yourself in time and space for a minute, and think of yourself as a part of the interwoven pattern of living and nonliving things. You are alive, so you are a part of the biosphere. You are holding a book made of paper, which came from wood, which came from living trees, another part of the biosphere. These trees, in turn, depended on and influenced the atmosphere, hydrosphere, and lithosphere during their lifetime. When the paper was processed, great quantities of water from the hydrosphere were used, and the process added other substances to the hydrosphere and atmosphere.

While you read, you are taking oxygen from the atmosphere and adding other gases to it.

You are sitting on something. It could be grass, a rock, or a chair. Whatever it is, it is resting, directly or indirectly, on the lithosphere and is related to all the spheres in their origins and relationship.

If you are lunching, munching, or drinking while you read, another whole series of interrelationships is involved. In fact, the pattern of interrelationships of one person on planet Earth at a certain moment could fill volumes.

You, this book, the air, and everything else that surrounds and touches you at this moment are composed of molecules of matter called compounds, made from elements, Earth's basic building blocks. Some of the compounds, such as the carbon dioxide that is forming in your body this very second, may be new. Others, such as the compounds that make up your brain cells or the pages of this book, may be several years old. Still others, such as the compounds that form the rocks of the lithosphere on which you, or your house, or your chair is resting, may be ancient, existing in exactly the same form that they did millions of years ago. Regardless of the age of these compounds, they are all composed of elements that have been

Meteoroid, meteor, and meteorite are three stages of the same object. Meteoroids are pieces of matter floating around in outer space. When they get close to Earth they are pulled by Earth's gravity and come crashing through the atmosphere.

We see them then as a streak of light and call them shooting stars or meteors. Generally meteors are completely consumed in their trip through the atmosphere and drift to Earth as meteoric dust. Once in a while a big meteor does not burn completely. When it lands it is called a meteorite.

This is the Willamette Meteorite, which was discovered in Oregon. The fourth-largest piece of matter from outer space found anywhere in the world, it is on display in The American Museum of Natural History in New York City.

Courtesy American Museum of Natural History

here since the solar system was formed four billion or more years ago.

Not only have these elements remained unchanged in their structure and behavior through all of Earth's history, most of them exist in exactly the same quantity that they did when Earth was first formed. A few minor additions and subtractions have occurred. Atoms of helium and hydrogen have wandered into the upper layers of the atmosphere, where Earth's gravity is not strong enough to hold them, and bounced into space. Each year meteoric dust and meteorites add some iron, nickel, and silicon compounds to the surface of the lithosphere. The atomic structure of a small number of the heavy radioactive elements, like radium, plutonium, and uranium, breaks down naturally. When this happens, infinitesimal quantities of these elements are changed, forming other elements, such as lead, helium, xenon, and radon.

These natural additions to and subtractions from Earth's supply of elements are minimal, and life has flourished on Earth for two billion or more years simply by using and reusing these wonderful building blocks. For that reason Earth is often compared to a spaceship. Everything needed for life is here. Earth may pick up a bit of matter in the form of meteoric dust from outer space, just as astronauts collect moon rocks, but there is no outside source for our life-giving substances. If we destroy them by making them into compounds that other living things cannot use, that we cannot reuse, or that Earth's recycling process cannot handle, we will destroy ourselves.

Physicists and chemists describe the atoms that make up the elements as bits of matter composed of a positively charged nucleus around which negatively charged particles called electrons travel at great speed. Protons are positively charged particles that are part of the nucleus, and there is one electron for every proton in the nucleus, so that an atom is neutral. The electrons travel in regular paths, similar to the orbits of the planets around the sun, except that the paths are spherical and are called shells. Each shell can hold a certain number of electrons. If the atom does not have enough electrons to fill its outermost shell, it joins with other atoms to form a molecule. A molecule of a compound is the smallest unit into which that substance can be divided and still have the properties (characteristics) of that compound.

Substances like the inert gases helium and xenon, whose atoms have enough electrons to fill their outermost shell, rarely, if ever, combine with any other substances.

The first shell of every atom can hold two electrons. An atom of hydrogen contains only one electron, so it has room for one more. Often two hydrogen atoms combine, which scientists call forming a chemical bond. A molecule of hydrogen contains two hydrogen atoms.

Diagram of a hydrogen atom (H) with one electron moving around the nucleus which contains one proton.

Chemists use symbols to write about elements. These are a kind of international shorthand. The symbol for hydrogen in any part of the world is H, and a molecule of hydrogen is represented by H_2. The number 2 indicates that there are two atoms of hydrogen.

Hydrogen combines with many other substances to form compounds. When it combines with oxygen, it forms water, a very important compound. An atom of oxygen has eight electrons. Two are on its first shell. That leaves six for the next shell. But the second shell of all atoms can hold eight electrons.Oxygen, like hydrogen, forms bonds

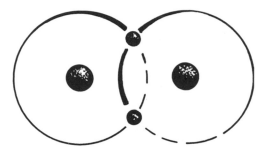

Diagram of a hydrogen molecule (H_2) consisting of two bonded hydrogen atoms.

easily with other elements whose outer shells do not have all the electrons they can hold. When two atoms of hydrogen and one of oxygen are heated, the two elements combine with an explosion and a flash of light and form a molecule of water. The formula for water is H_2O. This formula indicates that one molecule of water is made up of two atoms of hydrogen and one atom of oxygen. (The number 1 is not written in chemical formulas.)

Water is a compound. There are hundreds of thousands of kinds of compounds on Earth. Some, like table salt, are made of only two atoms; others, like chlorophyll, are made of hundreds of atoms; still others, like hemoglobin, consist of thousands of atoms in one molecule.

Table salt is sodium chloride, which is written NaCl. Na is the symbol for sodium, and Cl is the symbol for chlorine.

The first letter of a chemical symbol is always a capital, and if there is a second letter, it is always lower case. Most symbols are the first letter or the first letter plus one other letter of the element's name. Examples include carbon (C), oxygen (O), cobalt (Co), hydrogen (H), helium (He), chlorine (Cl), sulfur (S), magnesium (Mg), and nitrogen (N). A few symbols come from Latin or Greek words. For example, Fe, the symbol for iron, comes from the Latin word for iron, *ferrum*. Others, such as Na for sodium, tell something of the history of the element's use.

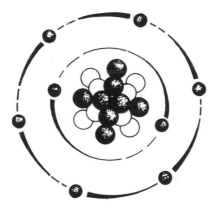

Diagram of an oxygen atom (O) with eight protons in the nucleus, two electrons on the inner shell, and six on the second shell.

Sodium always exists in compounds. No one ever saw the free (uncombined) element until 1807, when it was isolated by Sir Humphrey Davy, an English chemist who was knighted because of his research, which resulted in the discovery of seven different elements. For many centuries prior to the isolation of sodium, sodium compounds were well known and had many uses. A large deposit of one well-known sodium compound, sodium carbonate, was located in Spain, where it was called natron. The symbol Na recognizes the importance of this Spanish deposit of the compound.

The element sodium is a silvery-white metal, which combines so rapidly and violently with either oxygen or chlorine that it must be stored in some liquid or gas that

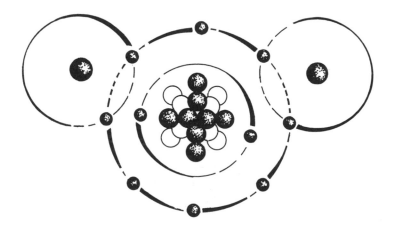

Diagram of a water molecule (H_2O). By sharing electrons, each hydrogen atom "borrows" an electron from the outer shell of the oxygen atom and completes its shell. At the same time the outer shell of the oxygen atom is filled with eight electrons, six from oxygen and two "borrowed" from hydrogen.

does not contain either of these elements, even as part of a compound. Usually it is stored in kerosine.

The element chlorine is a poisonous green gas. If a piece of the silvery metal sodium is put in a jar of the green gas chlorine in the sunlight, the two dangerous-to-life elements unite with an explosion and form a white crystalline substance that is essential to life—salt. If they are introduced into the same jar in the darkness, the reaction will be slower; there will not be an explosion, but the same cubical crystals of NaCl will be formed.

Because each compound has properties that are different from the elements that formed it, there is an immense variety of substances on Earth.

Chlorophyll and hemoglobin are two compounds that illustrate this variety. Both are protein substances whose main building blocks are carbon, oxygen, hydrogen, and nitrogen. But they differ greatly in the size of their molecules and in their behavior, as well as in the metal that gives them their special quality.

There are seven known kinds of chlorophyll. Each has a slightly different composition, but all are composed of the elements carbon, hydrogen, oxygen, nitrogen, and magnesium. The most common chlorophyll is called chlorophyll a. It is blue-green in color and is a medium-sized molecule containing 137 atoms. Its formula is $C_{55}H_{72}O_5N_4Mg$. Millions of these molecules would be needed to cover the period at the end of this sentence.

Hemoglobin, the red substance in human blood, is composed of much larger molecules, each consisting of 9,272 atoms. Its formula is $(C_{738}H_{1166}FeN_{203}O_{208}S_2)_4$.

These three compounds—salt, chlorophyll, and hemoglobin—are formed from varying quantities of nine different elements. These nine elements are some of Earth's most important building blocks, and are used over and over again. While there are ninety-two naturally occurring elements on Earth, plus another twelve that have been artificially formed in atomic reactors, they are not all equally important. Only nine elements make up 95 percent of the weight of the lithosphere. They are oxygen (O), silicon (Si), aluminum (Al), iron (Fe), calcium (Ca), sodium (Na), potassium (K), magnesium (Mg), and hydrogen (H). Three others, carbon (C), chlorine (Cl), and sulfur (S), are also widely distributed.

Except for sulfur and carbon these elements always exist as parts of compounds in the lithosphere. Carbon and sulfur usually occur as compounds also, but there are large deposits of free sulfur in areas of past and present volcanic activity, and two forms of pure carbon—graphite and diamonds—occur in scattered deposits.

Most of the gases in the atmosphere, however, are present as free elements. Earth's atmosphere has probably been much the same for the last two billion years. As recently as 1950, people believed that the ocean of air was so vast that nothing humankind did could influence it. Poisonous fumes like carbon monoxide (CO) and hydrogen sulfide (H_2S), or excess water vapor (H_2O), carbon dioxide (CO_2), and ozone (O_3) might be annoying locally, but it was thought that they would eventually be diluted and rendered harmless. Today the by-products of atomic explosions have been carried by air currents to every spot on the globe, and it is known that there are no geographical barriers to pollution.

More than 99.99 percent of the lower level of the atmosphere, which is the level that supports the biosphere, is composed of five gases in the following proportions:

nitrogen	N	78.084 percent
oxygen	O_2	20.946 percent
argon	Ar	.934 percent
carbon dioxide	CO_2	.034 percent
water vapor	H_2O	varying amounts

The balance, roughly .002 percent, consists of neon (Ne), helium (He), methane (CH_4), krypton (Kr), hydrogen (H_2), nitrous oxide (N_2O), xenon (Xe), radon (Rn), ozone (O_3), and dust.

Basically the hydrosphere consists of only two elements (H_2O), but because of the dissolving and carrying action of water, particles of many elements and compounds are dissolved and suspended in all bodies of water.

Finally, there is the biosphere, that wonderful thread of life which occupies all the other spheres. It, too, is composed of compounds formed from elements. Its survival is dependent on the arrangement of the building blocks into things that are often called food, shelter, air, water, and a favorable environment.

Diagram of a helium atom (He) with two electrons moving around the nucleus, which contains two protons.

Thirteen elements make up the human body in the following proportions by weight:

oxygen	(O_2)	64.5	percent
carbon	(C)	18	percent
hydrogen	(H_2)	10	percent
nitrogen	(N)	3	percent
calcium	(Ca)	2	percent
phosphorus	(P)	.9	percent
potassium	(K)	.35	percent
sulfur	(S)	.25	percent
sodium	(Na)	.15	percent
chlorine	(Cl)	.15	percent
magnesium	(Mg)	.05	percent
iodine	(I)	trace	
iron	(Fe)	trace	

All green plants require at least fifteen different elements. Some have special additional requirements. Altogether, forty different elements have been isolated from plants. Almost 99 percent of the body of a green plant is composed of carbon (C), hydrogen (H), and oxygen (O) by weight. In addition, phosphorus (P), sulfur (S), nitrogen (N), potassium (K), calcium (Ca), magnesium (Mg), and iron (Fe) are present in easily measured quantities. Finally, five elements occur in such small amounts that they can only be detected by the most delicate tests. They are called trace elements. If they are not available, plants die. The trace elements required by all flowering plants are boron (B), copper (Cu), manganese (Mn), molybdenum (Mo), and zinc (Zn).

A list of the elements in the bodies of beetles, fish, mushrooms, birds, and all other living creatures of the past and present would begin to be monotonous, for while a few trace elements would differ, the same basic materials would appear time and time again. It becomes obvious, then, that if plants and animals have been using huge quantities of carbon, hydrogen, and oxygen, and lesser quantities of nitrogen, sulfur, potassium, phosphorus, calcium, and other elements for two billion or more years, there would be none left for living things today if Earth were not an extremely efficient recycling plant. It should be equally obvious that if we hope to keep Earth habitable, we must live in harmony with these interacting interdependent spheres.

CHAPTER THREE

Water in Perpetual Motion

The variety of Earth's recycling processes is well illustrated by the most abundant of its compounds —water. The first water on Earth probably came from erupting volcanoes in the form of water vapor. As the vapor cooled, it condensed and fell as rain.

Water is unique in being able to exist in three states—solid, liquid, and gas—at the temperatures that occur naturally on Earth, and in being able to change easily from one state to another. This ability to change readily from one state to another is one of the most important properties of water.

It's easy to tell when maple syrup is being produced in a sap house, for as the water vapor pours out of the vents of the roof and hits the cold air, it forms a cloud of tiny droplets of water.
Courtesy Ken Williams, State of New Hampshire, Division of Economic Development

In 1943, lava suddenly rose from a cornfield in central Mexico, and a new volcano, Paricutín, was born. As this picture shows, basalt formed from lava on Paricutín is pitted with small pores, larger holes, and big fissures where water escaped, just like the volcanic rock from the many volcanoes that were erupting when Earth was young.

When it is very cold, water vapor freezes without turning into a liquid. Snow and frost are two forms of frozen water vapor. Here water under the snow evaporated during the day, then, as the temperature dropped at night, water vapor that had been trapped in the holes in the snow formed frost crystals on their edges.

In fact, if water were not constantly changing from a gas to a liquid and back to a gas, it would all be in the oceans or deep in the lithosphere today, for water responds to the pull of gravity by moving downward. Without rains to replenish them, streams and rivers would eventually empty into the oceans, and the land would be bare and dry.

Water evaporates continually wherever it comes in contact with the atmosphere. The rate at which it evaporates and the amount of evaporation are dependent on the temperature of the water, the temperature of the air, and the amount of water vapor already in the atmosphere.

Air can hold a set amount of water vapor at any given temperature. The warmer the air is, the more water vapor it can hold. The amount of water vapor in the air is generally called the relative humidity. A relative humidity of 50 percent means that the air contains half as much water vapor as it could hold at that particular temperature.

As water evaporates from the surface of the lithosphere and the hydrosphere, it is carried upward in warm air currents. Eventually, it reaches a level where the temperature is so low that the air cannot hold any more water vapor. (The relative humidity is 100 percent.) At this level the vapor condenses and forms droplets of water, which bunch together as clouds. Gradually, the droplets of water collect on dust particles until the particles be-

This stream with ice forming on its surface represents water as a solid, as a liquid, and as a gas. For although water vapor is invisible, the frost on the weeds and leaves above the stream records the fact that vapor has been escaping from the stream, while the snow is a record of frozen vapor in the clouds.

Frequently icicles are stained brown, orange, or other colors by the substances that the water dissolved or carried as it ran over the ground or percolated through cracks in rocks before it froze.

come so large that they can no longer withstand the pull of Earth's gravity and they fall as rain.

Often the rain falls far from the place where the water evaporated, for air currents in the form of wind are constantly moving from colder areas to warmer ones. In this way the water in the soil, the lakes, and the streams of the lithosphere is replenished.

The return of water to the lithosphere is only one contribution of this recycling operation, however. When water changes to water vapor, all the substances that the water was carrying are left behind. Regardless of how full of mud, minerals, living organisms, gases, sewage, or industrial wastes a body of water may be, the water vapor that escapes from it will be pure water.

Pure water in liquid form is extremely hard to find, for water dissolves and transports almost everything it contacts. For that reason water is called the universal solvent. Some things, such as salt, dissolve readily in water; others dissolve slowly.

As water evaporates from the oceans, it leaves behind the minerals that have been dissolved from the lithosphere and transported to the hydrosphere. In this way the concentration of minerals in the ocean has grown through the ages, with the result that today a cubic mile of ocean water contains an average of 89,500,000 tons of chlorine and 49,500,000 tons of sodium. This means that one cubic foot of ocean water contains almost two pounds of salt (NaCl). That same cubic foot of water contains roughly one and one-third ounces of magnesium (Mg), almost an ounce of sulfur (S), two fifths of an ounce each of calcium (Ca) and potassium (K), as well as the following sixteen elements in progressively lesser but measurable amounts: bromine (Br), carbon (C), strontium (Sr),

boron (B), fluorine (F), rubidium (Rb), iodine (I), barium (Ba), zinc (Zn), arsenic (As), copper (Cu), uranium (U), manganese (Mn), silver (Ag), lead (Pb), and gold (Au).

Thus the water cycle not only returns water to the land, it cleans and freshens it, so it can be used over and over again.

The water cycle is based on physical change. Regardless of whether water is in a solid, liquid, or gaseous state, its chemical composition is always H_2O. However, when water is a solid, its molecules are locked into each other and move relatively slowly; when it becomes a liquid, the molecules move more freely and rapidly, although they still contact each other; when water is a gas, each molecule moves extremely rapidly and may be totally independent of its neighbor. This increase in speed and motion requires energy.

Earth's water cycle functions because of the availability of the sun's energy. If there were less energy, all water would be locked in a solid state; if there were a great deal more, it would all be gas. Thus life-giving water is available because of Earth's ideal distance from the sun.

The ability of water to dissolve and transport, which makes it a great shaper of Earth, also makes it the pri-

The carrying power, the dissolving ability of water, the thrust of ice in cracks are all illustrated in this gorge cut by Van Campan's Brook in Pahaquarry Township, New Jersey. Notice the height of the original rocks on either side of the waterfall, the hole dissolved in rock at the foot of the falls, the loosened and transported materials.

Courtesy Herb Wiley, Jr.

GUARD CELLS

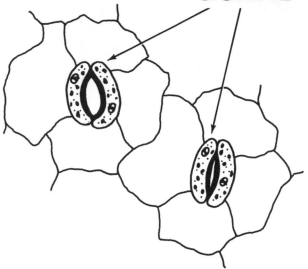

Each stoma on the edge and underside of a leaf is surrounded by two guard cells which control the size of the stomata by curving or straightening in response to the amount of air and water inside the leaf in relationship to the rate of photosynthesis and respiration taking place. This drawing shows stomata and guard cells as they look under a microscope with the stoma on the left open and the one on the right almost closed.

mary substance of life. The blood of animals and the sap of plants are actually water carrying dissolved minerals, gases, food, and other substances.

This fact is illustrated by the making of maple syrup, which is done by collecting the sap of maple trees and evaporating the water from it. Forty gallons of sap are used to obtain one gallon of syrup. About thirty-nine gallons of water are evaporated, and even then the syrup still contains water. If the syrup is heated until all the water has evaporated, about five pounds of the solid material called maple sugar remain. (Forty gallons of sap weigh about 320 pounds.)

Plants also contribute to the water cycle by a process known as transpiration. When plants take up water through their roots, it becomes a part of the sap and moves through the plant veins to every part of the plant. Surplus water escapes from tiny holes in the leaves called stomata. During the day this water evaporates immediately, but during cool nights, when the relative humidity around the plant is high, the water collects in little droplets at the stomata.

This process is called guttation. Drops of guttation are found at the tip of blades of grass and at regular intervals on the edge of many other leaves, looking like sparkling diamonds. It is different from dew because it comes from inside the plant and contains dissolved substances, whereas dew condenses from the air. If dew condenses on a clean leaf, it will be pure water.

You can test this by collecting drops of dew and drops of guttation on a piece of clean glass, permitting them to evaporate, and then examining the glass for traces of solid material.

Transpiration, which goes on only in living plants, moves huge quantities of water from the soil to the air. The process is exactly the same in a potted African violet in a city apartment as it is in a giant redwood tree in a forest. Botanists (scientists who study plants) and physicists (scientists who study the physical laws of the universe) are still carrying out experiments to try to discover exactly how transpiration moves water to the top of high trees.

Dew forms anywhere—even in deserts—when water vapor in the air condenses as it comes in contact with plants and other objects that cool rapidly at night.

Courtesy Arline Strong

Generally, only observant early risers are fortunate enough to see the sparkling drops of guttation that form when surplus water accumulates within the leaf during the night, for when the air around the plant warms, the drops evaporate.

Courtesy Arline Strong

A terrarium is a good place to see both transpiration and the water cycle working. If you examine the terrarium early in the morning, you will find that the glass on the top and possibly at the sides is covered with drops of water that condensed there during the night. Some of this water probably evaporated from the surface of the soil; some had been in the air as water vapor and condensed as dew when the temperature dropped, and the rest was transpired from the plants. Sometimes you will be able to see guttation in a terrarium, particularly if it contains grass, violets, strawberries, or *Impatiens*.

Industrial pollution dramatically changes the color of the river as well as the air in this Canadian town.

Courtesy Claire Murchie

The water cycle has been cleaning water and moving it over the lithosphere ever since the first water vapor condensed and collected in low places. Transpiration has been moving water from the soil to the air ever since green plants have been growing on land. But today there are water problems all over the world. Some places do not have enough water. Others are surrounded by water that cannot be used because it is polluted—it contains substances that make it unsuitable or unsafe for living things.

Most of the problems of water pollution are recent problems in terms of Earth's history. Before cities were built, people could safely use water from ponds, lakes, streams, and springs, or they could dig wells and tap the groundwater supplies. If they were careful to discard their dirty water and their body wastes a hundred feet or more from their water supply, disease-causing organisms died and big particles of dirt were filtered out by the time the water trickled through the soil and became a part of the water table. Fresh water in the form of rain replenished people's supplies.

As cities developed, a great deal of waste accumulated in a small space. The easiest way to get rid of it seemed to be to let water carry it away. So the waste products from the cities were dumped into the river. When there was only one city on a river, it could dump its waste downstream and take clean water from upstream, but as soon as two cities were located on the same river, the people of one city had to drink the other city's waste products. As industries developed, they, too, added pollutants to the water.

By the twentieth century, many streams and rivers were so contaminated with sewage and industrial waste that their water was no longer potable, or fit for drinking or cooking. Cities developed various techniques for treating water. They added chlorine to kill disease-causing organisms; they built settling basins so that mud and other substances picked up by water running off bare hills could settle out; they built filters to remove sediments from the water; they sprayed water into the air so oxygen could kill some kinds of bacteria and destroy bad odors; and they built reservoirs to collect and hold the water.

Some cities did more than treat their water. They looked for ways to stop polluting it.

Milwaukee, Wisconsin, developed a special process for treating its sewage. When the water leaves the processing plant, it is clear and free from disease-causing organisms and bad odors, and the solid material that remains is an odorless, disease-free substance that makes a good fertilizer. This material, called Milorganite, is packaged and sold in stores. In this way Milwaukee helps to pay for its sewage disposal; water is kept clean; and many gardens and lawns benefit from this natural fertilizer.

Some other cities are also beginning to use this technique. At Grand Canyon, Arizona, where water is very scarce, special pipes carry the clean water from this process to bathrooms where it is used in flush tanks. When the water is recycled in this way, it can be used over and over, just as water from the ocean is used and reused.

Many people feel that it is too expensive to process sewage so that water can be recycled and solid wastes do not pollute waterways. We are just beginning to learn that the cheap thing in dollars and cents may be more expensive in the end. If people are not healthy, if rivers are foul-smelling open sewers, if fish and birds die, if drinking water tastes and smells unpleasant, it's a bad bargain at any price.

Treating sewage is only one way to solve water problems. For years, when industries polluted water people said, "That's too bad, but it's the price of progress." Now we know that this kind of "progress" is another bad bargain. Laws are being passed requiring industries to find some other place to put their waste products.

Sometimes the industry discovers that the waste product can be put to some other use. As Lamont Cole, Pro-

fessor of Ecology at Cornell University and one of the leading ecologists of our times, has said, "Pollution is a natural resource out of place."

Soil in water is an excellent example of a natural resource out of place. Plants break the force of raindrops and slow down water, so it sinks into the ground instead of running over it. Plant roots hold the soil against the tug of running water. But water running over bare soil carries the loose particles, and if it moves fast, it digs gulleys and ditches. This wearing away of the land is called erosion. Erosion not only destroys fields, it fills lakes and reservoirs with solid materials. In 1933, the Tennessee Valley Authority was created to plan dams to manage the water in the Tennessee River and its tributaries. At that time, consultants from the United States Forest Service and the Soil Conservation Service were asked to plan the land use so that the lakes formed by the dams the engineers built would not quickly fill with silt but would remain useful for many years to come. Cities that plant trees around their reservoirs are protecting their water supplies in the same way.

As pollutants are removed from our waterways, more water will be available. In fact, if water were cleaned up, the natural recycling of water would provide enough for

A few years ago these two fields in Madison County, North Carolina, were equally gullied. The owner of the field at the left entered the TVA test demonstration program and by proper fertilization now produces a crop of pasture grasses and clovers that check erosion and provide food for cattle. Meanwhile in the neglected field each rain carries more soil onto the road and into waterways. ("A natural resource out of place.")

Courtesy Tennessee Valley Authority

most places. There are areas of the world, however, where rainfall is too sparce to supply enough water for cities or agriculture.

More and more people are looking toward the biggest of all reservoirs, the oceans of the world. If a cheap and simple way for evaporating and condensing ocean water could be devised, then the dry desert lands could be used for cities and farms.

Evaporating and condensing water is called distillation. Water is distilled on ships, and sometimes it is distilled by industries that need pure water to make their products or for scientific experiments. You could distill water in your kitchen by holding a cold plate over a pan of boiling water and catching the drops that condensed there. But it would be very expensive to produce clean water in that way. In fact, it is the cost of the fuel that would be needed to distill a great deal of water that makes distillation impractical. Of course, producing that much heat would create another pollution problem.

At the University of Arizona, Dr. Aden Meinel and his wife, Marjorie Meinel, have been doing research in trapping the sun's energy in specially constructed reflectors called solar farms. Energy caught in this way could be used to distill large quantities of water and to produce electricity without polluting the air.

Solar farms would only work, however, where there are many days of bright sunlight and a lot of open space. The southwestern United States has this kind of space and sunshine. So have the Sahara desert and the desert of Peru.

Another experiment is being carried out in Peru. Water from the Pacific Ocean is being piped into shallow basins in the desert. Specially constructed tents of clear plastic cover the basins, but permit the sun's rays to warm the water and evaporate it. As the fresh water condenses on the roof of the tents, special gutters collect it and conduct it to irrigation ditches. If this can be done on a large scale, Peru will be able to produce much more food by putting the sun and the water cycle to work.

Parts of the world that are farther from the equator cannot use the sun's energy as easily as this, but as people everywhere learn more about the natural recycling processes that take place on Earth and try to become a part of the pattern, many of the problems that have arisen from humankind's ignorance or disregarding of natural processes can be solved.

Dr. William Beauchamp of the Department of Optical Sciences at the University of Arizona is demonstrating a solar-power test model which is used to convert solar energy into electrical energy.

Courtesy George Kew

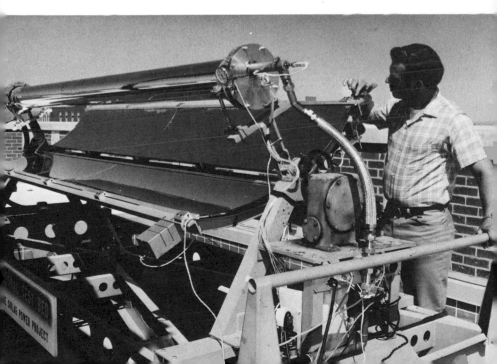

CHAPTER FOUR

Green Plants—Energy Traps

Have you ever stood and looked at new babies and thought of all the growing they would be doing? Their organs will have to increase in size. They will have to grow teeth and hair. They will need to grow a lot of new skin and longer, bigger bones and bigger, stronger muscles. And, of course, they will have to add more brain cells. In other words, they have a big construction job ahead of them.

Of course, you have already gone through this. In theory you should be able to tell exactly how it is done. But if someone asked you, "How did you know to add calcium to your bones and iron to the hemoglobin in your blood?

Many of Earth's cycles can be studied in terrariums that can be made in almost any transparent container. Notice the sand on the bottom of this one to provide drainage.
Courtesy Wendy Holmes, Wave Hill Center for Environmental Studies

How did you know to grow hair instead of feathers on your head? Why did you grow your particular shade of skin color?" you'd realize that there are many things that we do not control.

Biologically we belong to the animal part of the biosphere, and when we start life, we inherit a built-in programming from our parents, as insects do from theirs and plants do from theirs. All living things inherit their particular plan, which not only determines what they will be but what building blocks they will need and how these building blocks will be assembled. This built-in programming is part of the special quality of being alive.

All living things are basically made of protein substances called protoplasm. Four building blocks—oxygen, hydrogen, carbon, and nitrogen—make up almost 99 percent of this living tissue.

Oxygen, hydrogen, and nitrogen exist in an uncombined state in the atmosphere. Pure carbon, too, exists as graphite and diamonds. But no living thing can take these elements and build protoplasm with them. To be used by

Because the sun is so bright, the corona (hot gases) that surrounds it can only be studied and photographed during an eclipse, when the disk of the sun is covered by the moon's shadow.
This picture is composed of two photographs, one taken during the eclipse of March 7, 1970, and the other taken with an X-ray exposure through a telescope mounted on a rocket. As scientists study photographs taken by these special techniques, they learn more about Earth's source of energy.
Courtesy of the Solar Physics Group American Science and Engineering, Cambridge, Massachusetts, and the High Altitude Observatory, Boulder, Colorado

living things, all the elements must be in the form of specially formed compounds.

The formation and breakdown of compounds is called chemical change, and chemical changes always involve a transfer of energy. Earth's energy comes from the sun. Two forms of solar energy are heat and light.

Green plants are the only living things that can trap and use solar energy. Scientists believe that when light strikes the molecules of chlorophyll, the solar energy is transferred to the outer electrons of the chlorophyll molecules. These electrons then move in bigger orbits, and as they bump into other electrons, they transfer to them the energy that they picked up from the sunlight.

When this energy is transferred to molecules of water inside the leaf, each water molecule breaks into atoms of hydrogen and oxygen. Two atoms of oxygen combine with each other to form a molecule of oxygen. These molecules of oxygen move through the spaces in the leaf, and when they come to the stomata, they escape into the atmosphere.

The hydrogen atoms take a different path. They stay inside the plant leaves. By a complicated series of chemical changes that can take place only in the natural laboratory inside a plant leaf, these hydrogen atoms combine with molecules of carbon dioxide to form sugar. This process of using solar energy to produce food is called photosynthesis. The word *photosynthesis* comes from two Greek words meaning "to put together with light."

The sugar produced by photosynthesis is called glucose. Its chemical formula is $C_6H_{12}O_6$. Within the leaf, glucose may be changed to other kinds of sugar with a slightly different formula, or it may be changed to starch, fat, or proteins.

Much of the food that the plant produces is stored for future use, but some of the glucose is used immediately to release the energy needed to make new compounds for growth, reproduction, and other life processes. To release this energy, the glucose combines with oxygen. One molecule of glucose combines with six molecules of oxygen. The plant does not plan or direct this process any more than you do when you use your food. In fact, the process is identical. Chemists tell the story by writing the equation: $C_6H_{12}O_6 + 6O_2 \rightarrow 6CO_2 + 6H_2O$. If you count the

$$C_6H_{12}O_6 + 6O_2 \longrightarrow 6CO_2 + 6H_2O$$

To add the atoms of two or more compounds, we multiply the number of molecules of a compound by the number of atoms of each element. Since sugar has no number in front of it, we are talking about one molecule of this compound, so:

$1 \times 6 = 6$ atoms of carbon
$1 \times 12 = 12$ atoms of hydrogen
$1 \times 6 = 6$ atoms of oxygen

If we are going to add 6 molecules of oxygen to it and each molecule contains two atoms, we multiply:

$6 \times 2 = 12$ atoms of oxygen

Thus the left side of the equation has 6 atoms of carbon, 12 atoms of hydrogen, and 18 atoms of oxygen: $6 + 12 = 18$. How does this compare with the number of atoms on the right side? (Remember one is never written in an equation. Any symbol without a number after it is always one atom, just as any molecule without a number in front of it is one molecule.)

atoms of each element, you will find that there is exactly the same amount of oxygen, hydrogen, and carbon on each side of the equation. These building blocks are used to form different molecules, but not one atom of the elements is lost in the process.

When a substance combines with oxygen, the process is called oxidation, and the substance is oxidized. When glucose is oxidized, it forms carbon dioxide and water, and the energy that the plant used in forming the glucose is released in the form of heat. This energy may be used to initiate other chemical changes. In your body some of the heat energy from the oxidation of food keeps your body temperature at 98.6° F. (37.0° C.). Some of the rest of the energy enables other chemical changes to take place; thus the new compounds that make up your body tissues are formed.

All life processes involve chemical changes. In the biosphere the sun's energy, which is trapped by chlorophyll, provides the energy for these chemical changes. At the same time that it is trapping the sun's energy, chlorophyll enables compounds to form that plants use to build tissues and carry out life functions. Chemists call chlorophyll a catalyst. Catalysts are substances that either speed up a chemical reaction or cause it to take place without undergoing chemical changes themselves.

Without chlorophyll, the catalyst that traps solar en-

These mushrooms are non-green plants that live on dead wood, leaves, and other plant parts in the soil, taking in the sugars, starches, proteins and/or fats that the green plants produced when they were alive and oxidizing them for energy, growth, and reproduction.

ergy, the biosphere could not exist. In fact, the process of photosynthesis provides food for all living things. It also is a tremendous recycling operation.

Plants are continually using carbon dioxide and water to produce glucose and other food products. Plants and animals are constantly oxidizing food and releasing carbon dioxide and water.

Photosynthesis is often referred to as the oxygen–carbon dioxide cycle because plants release oxygen at the same time as they use carbon dioxide in producing food. Sometimes people think plants breathe out oxygen and take in carbon dioxide, and animals breathe out carbon dioxide. *This is not correct.* Plants and animals both use oxygen to oxidize their food, and both produce carbon dioxide and water when they do this.

Plants take oxygen from the air when there is no free oxygen inside the leaf left over from the process of photosynthesis. The oxygen combines with food to make CO_2 and H_2O. When the sun is shining, they use the carbon dioxide and water that are released to make more food, but at night and on dark days, carbon dioxide and the water vapor are added to the atmosphere. In other words, plants carry out the same processes as animals in using the food they produce.

The process of oxidizing glucose and other food materials in living tissues is called respiration. The process of respiration in green plants is exactly the same as the process that takes place in your body, in other animals, and in nongreen plants like mushrooms and bacteria.

Photosynthesis and respiration are two interwoven processes. Without this constant changing of carbon dioxide and water into sugar and oxygen and vice versa, the biosphere could not exist, for the building blocks would quickly be unavailable if they remained in one form or the other.

Respiration and photosynthesis use exactly the same building blocks. If plants made only as much food as they needed for immediate use, they would not increase the amount of free oxygen in the atmosphere. It is only the healthy, growing plants which are producing surplus food that release the oxygen needed to replenish the supply in the atmosphere. That supply is constantly being depleted by all the oxidation taking place.

You breathe an average of sixteen to seventeen times per minute, and in the process of breathing you take air into cavities in your lungs. There the oxygen molecules pass through the thin lung membranes and combine with hemoglobin to form an unstable compound called oxyhemoglobin. An unstable compound is one that breaks down easily. As the oxyhemoglobin is carried through your body, it breaks down into oxygen and hemoglobin whenever it comes to a place where there is not very much oxygen in your tissues.

While oxygen from the air enters your bloodstream, surplus carbon dioxide and water vapor, which were formed in your tissues when your food was oxidized, pass through the thin lung membranes to the air pockets and are exhaled.

Each day your body uses an average of three pounds of oxygen and returns it to the air in the form of carbon dioxide and water. If you multiply three pounds by the population of the place where you live, by the population of a major city like New York, Tokyo, Mexico City, Paris, or Los Angeles, and finally by the world population of more than three billion people, you will have some idea of the vast amount of oxygen used every day by people. Then

think of all the animals—the wild and domestic mammals, the birds, reptiles, and fish, amphibians as well as the millions of small animals such as oysters, insects, shrimps, spiders, and protozoa—all using oxygen from the air in varying amounts. Some animals, like horses and elephants, individually need more than people; others may need very little individually but exist in huge populations. Remember that plants, too, both green and nongreen ones, are using oxygen constantly in amounts varying with their size and activity. Even though the atmosphere in which

The pond, the fields, and the surrounding countryside are all a part of this ecosystem which supports more than a thousand wild geese in southeastern Pennsylvania.

Courtesy Russell Yeager

the biosphere exists is about 21 percent oxygen, the daily drain on this supply would quickly deplete it if green plants were not constantly taking in the two compounds —carbon dioxide and water—that all living things produce in the process of respiration, and were not reconverting them to food and free oxygen.

Frequently someone looks at a terrarium and says, "How does it get air?" Plants in a tightly sealed terrarium can grow successfully for a long time because they are such efficient recyclers. Eventually, however, the plants will die as carbon dioxide and water become parts of larger molecules in the plants' tissues, unless fresh air and water are introduced. Exactly how many months or years plants can grow in a sealed terrarium depends on the size of the container, on the size of the plants, and on their rate of growth.

All other things being equal, the plants in the terrarium will last longer if it also houses some small animals such as earthworms, a snail, a slug, or a cricket. If you want to see the carbon dioxide–water/glucose–oxygen cycle work, the animals you introduce must be ones that live on dead or living plant material, so that you will not have to take the lid off to introduce food.

Keeping an aquarium is an equally good way to observe this recycling process. The amount of oxygen dissolved in water is one of the most important factors determining whether or not fish can live in it. As the water enters the fish's mouth and passes out over its gills, oxygen dissolved in the water moves through the thin gill membranes into the fish's blood. There it forms oxyhemoglobin in exactly the same way that it does in your body and the bodies of other red-blooded animals. As the oxygen is used in the fish's body to oxidize its food, it is expelled as carbon dioxide and water.

If there are plants in the water, they use carbon dioxide and water to form food and release oxygen, which is one way that oxygen gets into water. Oxygen and other gases in air also enter the water through its surface. Sometimes people use electric pumps to pump air into their aquariums, either because they are trying to rear fish that live in moving water, or because they do not want to take the time or trouble to make a balanced aquarium.

In a balanced aquarium there are enough green plants to produce extra food and thus to release extra oxygen to keep the fish and snails supplied with the oxygen needed to oxidize their food. They, in turn, provide enough carbon dioxide for the green plants to produce food and oxygen for themselves as well as to produce surplus food—thus releasing the oxygen that fish and snails can use . . . and so on ad infinitum! Building a balanced aquarium is a challenge, but this balance is exactly what happens in ponds, lakes, and oceans that successfully support animal life.

A terrarium with animals in it or a balanced aquarium is really a small ecosystem. Earth is made up of millions of ecosystems. Each ecosystem includes parts of the lithosphere, hydrosphere, atmosphere, and biosphere that function as a unit.

All of Earth's ecosystems taken together are called the ecosphere. The ecosphere is that narrow band, less than fifteen miles wide, of hydrosphere, lithosphere, and atmosphere where life exists.

The biosphere of any ecosystem is always made up of green plants and of animals that feed on green plants. It may also have animals that feed on animals that feed on green plants.

An ecosystem may be as small or smaller than a tide pool or as large as a forest, but it always contains all the requirements for life for everything in it; and since living things are constantly using the building blocks of life and forming new compounds with them, all ecosystems contain all of Earth's recycling mechanisms.

CHAPTER FIVE

The Undercover Crowd

In addition to green plants and animals, every ecosystem contains millions of undercover workers in the form of such plants as fungi, blue-green algae, and bacteria. As important as photosynthesis and respiration are in making the sun's energy available to both plants and animals, they do not provide all of the necessary building blocks for life.

Protoplasm, which all living things are composed of, consists almost entirely of protein substances and water.

As this shelf fungus uses the food that was stored long before by the living tree, it helps break the wood down into its original substances: carbon dioxide, water, and various minerals. After it dies, bacteria will convert its tissues into other small molecules including nitrogen compounds.

Courtesy Herb Wiley, Jr.

Proteins always contain nitrogen. Although free nitrogen makes up approximately 78 percent of the atmosphere, neither green plants nor animals can use it in this form. To build protein, plants and animals need nitrogen compounds. These compounds are produced by several kinds of bacteria.

One kind of these bacteria lives in small nodules on the roots of members of the legume family, such as beans or peas. The scientific name for this group of bacteria is *Rhizobium*, a word which comes from two Greek words: *Rhizo*, root, and *bio*, life. These bacteria, which live on roots, are truly life-giving organisms, for they are unique in being able to take free nitrogen from the air and combine it with hydrogen to form ammonia (NH_3). The legume on whose roots *Rhizobium* lives provides the bacteria with food in the form of sugar and starch, and at the same time it uses some of the ammonia produced by the bacteria to make amino acids.

Legumes do this by combining the ammonia (NH_3) with carbon-hydrogen-oxygen compounds that started out as the sugar ($C_6H_{12}O_6$) that they produced by photosynthesis. Twenty-four different amino acids are known. Each consists of carbon, hydrogen, oxygen, and nitrogen atoms, which are strung together in different quantities and in different patterns. One amino acid is called glycine. Its chemical formula is $C_2H_5O_2N$. Another is alanine,

Each nodule on the root of legumes like this clover plant contains thousands of *Rhizobium*.

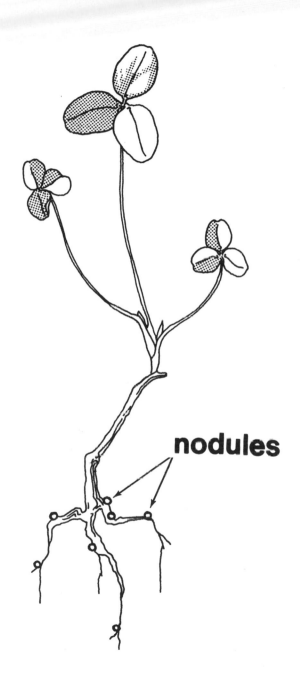

nodules

$C_3H_7O_2N$. A few amino acids also contain atoms of sulfur or potassium.

Once the amino acids are formed, proteins are produced in the plant through the combining of many of the same and/or different amino-acid molecules with each other and with molecules of other substances to form a single large molecule. Remember chlorophyll $(C_{55}H_{72}O_5N_4Mg)$ or hemoglobin $(C_{738}H_{1166}FeN_{203}O_{208}S_2)_4$? The large hemoglobin molecule is produced in human bodies through the rearranging of amino acids and an iron compound that comes from food.

As we consider this use of nitrogen atoms, it appears to be a passing-on and reusing of materials from one organism to the next. The *Rhizobium* produces ammonia and passes it on to the legume. Beans, peas, peanuts, alfalfa, clovers, and vetches are examples of legumes. The legume, such as a clover, uses the ammonia to make small molecules of amino acids, then combines these to make proteins. A cow eats the clover and digests the proteins, breaking them down into the amino acids from which they were formed, then rearranging them to make new proteins.

When you drink milk or eat a hamburger, your body breaks these proteins down into the original amino acids and recombines them to make the many kinds of proteins that make up the human body. (There are about 100,000 different proteins in your body.)

This appears to be a straight-line relationship, but a straight-line relationship that dealt with Earth's building blocks would eventually tie everything up in unavailable compounds, and life would grind to a halt. So we have to ask, "What next?"

Actually, several different things happen.

As time goes on protoplasm wears out, and the worn-out protein is removed by the body and replaced by new protein. The old protein is broken down into smaller compounds. Some of these small compounds may be re-used in building new tissues; others are discarded as waste products.

In animals this waste product is urea $(CO(NH_2)_2)$. Urine is composed of water (the universal solvent), urea, and other salts and minerals that the body does not need. Not only is urea produced from worn-out protoplasm, but whenever an animal eats more protein than it needs, its body digests the protein, oxidizes the carbon and hydrogen for energy, and discards the nitrogen as urea.

Some kinds of bacteria that live in soil use urea to produce ammonia. Still other soil bacteria take ammonia and change it to nitrites. Nitrites are compounds that contain one atom of nitrogen and two of oxygen (NO_2). Another kind of bacteria takes nitrites and forms nitrates, which are the main source of nitrogen for green plants, since they cannot use nitrites. Nitrates contain one atom of nitrogen and three of oxygen (NO_3). Some common nitrates are potassium nitrate (KNO_3) and sodium nitrate $(NaNO_3)$.

Nitrates dissolved in water enter the roots of plants. The plant uses them to make amino acids, and the cycle is complete.

There are many variations on this cycle. When plants and animals have lived for the normal period of time for their genus, or when they have lost the ability to replace worn-out tissues or grow, due to the failure to obtain the right substances for building protoplasm or the presence of some disease organism, they die. When plants and animals die, their bodies are broken down into small particles by a variety of other organisms that use them for food. In the process these organisms play a tremendously important recycling role.

These cycles, involving chemicals that are passed back and forth between living things and their environment through the actions of living things, are called biogeochemical cycles. Geochemistry means "Earth chemistry." It is the study of the chemistry of the nonliving components of planet Earth.

It would be impossible to list all of the plants and animals that are a part of the various biogeochemical cycles. You can observe some of these organisms at work if you can find a dead log or even a piece of board lying on the ground. If the wood is still tough and firm, it may contain wood-eating beetles, carpenter ants, and termites. If it has been around for a long time and has served as food for many of these animals, it may be the home and food

As one group of organisms finishes its work changing the dead-tree environment, it is replaced by another group that is adapted to the new conditions. We call this orderly change and replacement succession. This is a chart of a log that has been around a long time. The outer layers at this point look more like soil than wood, the next layers are soft and full of tunnels. Only the center is capable of supporting the organisms that feed on tough wood.

Once these organisms were on the outer layers. As they broke down the wood fibers, they moved inward. Eventually there will be no tough wood left, and they will have to move or die, but they will be succeeded by organisms from the second group and finally by the organisms that live in the soft powdery material—until eventually the log has been entirely broken down into the small molecules that green plants need to grow, and the log succession will be complete.

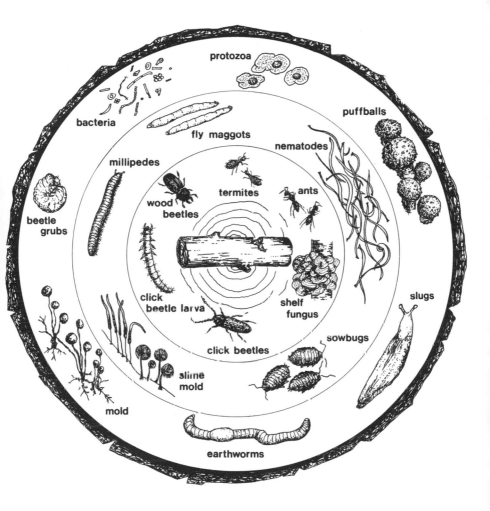

protozoa

bacteria

fly maggots

puffballs

millipedes

nematodes

beetle grubs

wood beetles

termites

ants

click beetle larva

shelf fungus

slugs

click beetles

sowbugs

slime mold

mold

earthworms

supply for millipedes, earthworms, sowbugs, pillbugs, snails, and slugs.

You may find white threads running through the soft powdery remains of the wood. These are fungus plants that use the wood for a food supply, and they send up mushrooms, puffballs, and other reproductive structures when food and water are abundant and the time is right.

You won't see bacteria, but several kinds will probably be there. Some will be feeding on the small particles of dead wood; others may be using the excrement of the animals feeding on the wood; still others may specialize in dead bodies of animals, such as beetles and earthworms that have met with accidents or completed their life-span. Whatever their food, all these bacteria are carrying out the last steps of converting the wood to its original materials —carbon dioxide, water, ammonia, free nitrogen, minerals, salts, and other small molecules.

The idea that things like carpenter ants, termites, and bacteria are important to the continuation of life on Earth is hard for many persons to accept. After all, when termites get in the beams of a porch, the porch collapses; or if they get into a library, they ruin whole shelves of books. And everyone knows that bacteria cause disease!

All of this is true. Certainly when termites and carpenter ants feed on dead wood that humankind is using for homes, furniture, and other products, they are destructive and must be removed, but most of the termites and ants of the world live outdoors, where they contribute to the success of the ecosphere.

When the French scientist Louis Pasteur did his famous studies in the 1850's, the world first learned about bacteria —the microscopic plants that cannot make their own food but must obtain it from the tissues of dead or living plants and animals. Of course, the first bacteria that were discovered were ones that spoiled food or caused illness or death in plants and animals, and it took a long time for people to realize that there are many more kinds of useful bacteria than harmful ones. Even today, we hear much more about the disease-causing bacteria called germs than about the many kinds that are essential for life to continue on planet Earth.

In addition to the different kinds of nitrogen-fixing bacteria and the bacteria that cause decay, there are other bacteria that rearrange sulfur and iron compounds into molecules that plants can use.

Even though the role of bacteria and other small organisms in biogeochemical cycles is a recent discovery, people have known for thousands of years that plants take things out of the soil, and that returning plant materials or animal products to the soil improves crops.

Sometimes people chopped up the stalks of plants such as corn, tomatoes, and pumpkins and added them to the soil. Indians in Massachusetts taught the English settlers to bury a dead fish in every hill of corn. Sometimes wood ashes from fireplaces were added to the soil, as well as manure from stables. When people fertilize the soil, they are putting back the compounds that plants need to grow after the previous crop has removed them. Sometimes human excrement was used as fertilizer. Human excrement is just as rich in the compounds that plants need as animal manure is, but it is a dangerous thing to use unless, like Milorganite, it has been treated to kill all organisms that cause human disease.

Even though the importance of fertilizer has been known for hundreds or thousands of years, in the first 150 years of United States history there were many persons

who cleared and burned forests and grasslands and raised crops without returning anything to the soil because they could always move farther west to new land. This was a very destructive practice, and some of the land that was mistreated in this way has never recovered.

A few leaders urged people to care for the land. One person who wrote and lectured about care of the land was Thomas Jefferson, the third president of the United States. He urged people to grow different crops in each field every year. Many people were growing nothing but cotton year after year, and others were growing nothing but tobacco.

Tobacco uses great quantities of nitrogen compounds, as well as large amounts of minerals from the soil. When tobacco is harvested, there is no plant part to return to the soil.

If a field was used for a tobacco crop one year and a clover crop the next, the bacteria on the clover roots would restore nitrogen compounds to the soil. Then other crops could be planted that would hold the soil better and remove less nitrogen than tobacco; finally, the farmer might want to use the field for tobacco again for one year. Planting a different crop in a field each year for a series of years is called crop rotation. It gives Earth's recycling processes time to restore some of the necessary compounds.

About one hundred years after Thomas Jefferson tried to persuade farmers to use crop rotation, George Washing-

When these cornstalks are plowed under, they will return some of the minerals to the soil that the corn plants used the previous year.
Courtesy Russell Yeager

ton Carver, a black scientist who was one of the out-
standing research botanists of all times, urged people to
raise peanuts instead of cotton or tobacco, or to alternate
peanuts or sweet potatoes with these two soil-robbing
crops.

Peanuts are legumes. While the peanut plants are producing the underground seeds that we call peanuts,
Rhizobium on the roots are producing ammonia. If the
plants are plowed under after the peanuts are harvested,
the field will often contain more available compounds for
plant growth than it did when the field was planted.

To provide a market for peanuts and to make it profit-
able for farmers to grow this crop, Dr. Carver spent many
years doing research and developed over one hundred
uses for peanuts. He also developed many uses for sweet
potatoes. Sweet potatoes do not produce nitrogen compounds, but they take less from the soil than tobacco and
cotton. Furthermore, their creeping pattern and many
roots hold the soil in place, make water move slowly, and
prevent the minerals from eroding the sandy soils of the
coastal plain of the southern United States. After the
sweet potatoes have been harvested, the vines can be
plowed into the soil.

In the early part of the twentieth century, scientists de-

Farmers can produce excellent crops year after year without damaging their soil by using manure to provide organic substances and
nitrogen compounds and only applying commercial fertilizers when
soil tests indicate that certain elements or compounds are missing,
and by following a program of crop rotation while using some
soil-holding plants like these sweet potatoes.

Courtesy Tennessee Valley Authority

When energy production results in air pollution, the farmer pays for his fertilizers and other products twice—once when he purchases them, and again in lower income from poorer crops because unburned carbon (smoke) clogs stomata of leaves and cuts down sunlight for photosynthesis while poisonous gases slow down plant growth and even kill crops.

Courtesy Everett Murchie

veloped techniques for making nitrogen compounds by passing great charges of electricity through liquid nitrogen and hydrogen or nitrogen and oxygen. Many farmers started to use these commercial fertilizers.

Commercial fertilizers provide plants with the same essential compounds that the soil bacteria produce. There is a difference in the field's fertility, however, for when the compounds are produced by biogeochemical recyclers breaking down the waste products and dead bodies of plants and animals, the small particles of these substances improve the texture of the soil and serve as sponges in holding water.

The use of electrically produced fertilizers creates several other problems. Generating electricity by anything but water power or solar energy always results in great quantities of air or water pollution. The man-made electricity that is used for power to produce chemical compounds either creates pollutants such as hydrogen sulfide, sulfur dioxide, or carbon monoxide or releases potentially dangerous radioactive substances.

At the same time, when commercial fertilizer is used, the natural products of life in the form of sewage, garbage, and manure pile up and create health and disposal problems. Most people do not realize what great quantities of waste materials are produced by humans and domestic animals unless something unusual happens.

In 1968 the garbage collectors in Toronto, Canada, went on strike. Garbage began to pile up in homes and on streets. It piled up in the Perth Avenue School, too. It created a real pollution problem until someone began to think of the garbage as "a natural resource out of place." Then classes in the Perth Avenue School began to carry out research on ways to use this organic material. The students learned about composting. Today the school is still using its garbage to produce compost for its own gardens and for the community.

Compost is a mixture of decaying organic substances (dead plant and animal materials) that is used in fertilizing the land. It is produced by cooperating with the biogeochemical recyclers.

Under some circumstances garbage may be buried in

Fruit and vegetable peelings, eggshells, coffee grounds, tea leaves, and tea bags (remove any plastic tags) can all go into a compost heap or be turned under the soil as organic fertilizer. If the yard or compost heap is dog-proof, bones and shellfish parts are excellent additions.

a yard or garden without being composted; earthworms, bacteria, and other organisms will break it down and make it available for plant use. But where there is a lot of organic material to be disposed of, it is better to make a compost heap. A compost heap may be protected from dogs, rats, and other vertebrate animals by building special boxes or pits. In compost piles or pits the temperature, moisture, and chemical composition can be modified to produce ideal conditions for bacteria and other organisms that carry on the recycling process.

As more and more gardeners are discovering that compost produces good crops and healthy lawns, more and more people are asking, "How do I compost my organic wastes?" Because of this, many state departments of conservation or environmental science are publishing instruc-

This is a Jersey City garden that kitchen compost helped to build. Four years before this photograph was taken, this yard was a poorly drained weed patch. In the intervening years Bob Russell has buried almost a ton of kitchen compost and watched earthworms and other recyclers convert it to humus.

This compost heap is composed entirely of grass clippings, remains of garden annuals, and leaves of trees that Doris Phillips of Glastonbury, Connecticut, rakes off the lawn. In ten years' time the interior has become a rich brown material, perfect for adding to flower beds, lawn, and garden. (Notice the band of this organic soil between the fresh leaves in the foreground and the partially decomposed grass and leaves on top.)

tions. The World Health Organization of the United Nations has published a book written by Harold B. Gotaas called *Composting, Sanitary Disposal and Reclamation of Organic Wastes.*

This book not only contains information on composting for small gardens and farms but provides information on large-scale composting of both garbage and sewage. This is extremely important, for when big cities learn to put the natural biogeochemical recycling processes to work, other parts of the interwoven spheres of planet Earth will profit. Waterways will not be polluted by sewage. Farmland will be enriched. Costs may be reduced by the sale of organic materials like Milorganite, so that tax money can be diverted to solve other problems. Worn-out or damaged land may be reclaimed, as it is in parts of Illinois, where sludge from biogeochemically treated sewage from the city of Chicago is being piped onto mine dumps and other wasteland in order to cover the bare mineral soil and turn deserts into forests and parkland.

This corn crop is growing on what was formerly an unsightly wasteland left behind by strip mining. This is one of the areas that had been leveled and filled with sludge from the completely processed sewage from the city of Chicago. Giant spray pumps in the background are spreading sludge in an adjacent area.

Courtesy The Metropolitan Sanitary District of Greater Chicago

CHAPTER SIX

Candy Bars and Relativity

It is impossible to talk about Earth's building blocks and the things that happen to them without talking about energy. Anything that involves changes in position, arrangement, combinations, or construction of the elements involves energy. The water cycle, photosynthesis, respiration, and biogeochemical recycling are all energy stories.

Generally the chemical and physical changes that took place when coal was formed two or three hundred million years ago destroyed all of the characteristics of the plants that were converted into coal; but once in a while a few specimens like these fern fronds from the anthracite coal regions of Pennsylvania left an imprint. These fossil records help us understand some of Earth's recycling activities and energy stories of the long-ago past.

Courtesy Pennsylvania Historical and Museum Commission

Energy is one of those things that everyone talks about but no one fully understands. If you stood on the street corner as a roving reporter and asked the first ten persons who came by to tell you about energy, you would receive a wide variety of responses. Most of the answers would be in terms of people's activities. For instance, people might say:

"Energy? That's what that terrible two-year-old I baby-sit for has too much of."

"I don't have the energy I had when I was forty."

"You need lots of energy to play football."

"A candy bar gives you quick energy."

"Energy? We study that in physics. It's the power to do work—to move things."

"Einstein was the man who finally discovered what energy is. He said, '$E = MC^2$.' That means energy equals mass times the speed of light squared."

It may seem that these statements have little in common: a candy bar and $E = MC^2$. If we start with the candy bar, we are talking about a food that is mostly sugar. Oxidized, it releases energy in the form of heat, which is used to initiate other chemical changes that eventually result in motion in a two-year-old, a football player, an octogenarian, or any other organism that uses it.

If we start with the sugar in the candy bar, we can say it came from a plant whose green leaves trapped the sun's energy and used it in forming a chemical compound which later released energy when it was oxidized.

This description tells us about energy moving from one place to another, but it does not really say what energy is. The physics student defined energy when he said it was the power to move things: to move yourself down the street, to throw a football, to run away from a baby-sitter.

The moving truck, the flowing water, and the running boy are each expressions of energy at one moment in time. Each also is the product of an infinite number of energy exchanges that took place naturally in the case of the boy and the water, and under human-kind's direction in the case of the truck.

Courtesy Chelsea Clinton News

These are obvious forms of movement; but the basic movement is what occurs in green leaves when energy from the sun agitates the outer electrons of the chlorophyll molecules and sets a whole series of molecular motions going.

Tracing energy back to its beginnings always leads to the sun. Gasoline comes from petroleum, which is the fossil remains of microscopic plants and microscopic animals that fed on those plants millions and millions of years ago. Coal and peat are the fossil remains of larger plants that flourished in swamps about 300 million years ago. Natural gas arises from deposits of plant materials. All go

Some of the molecules of water flowing over Bridal Veil Falls in Yosemite may have been part of a sugar molecule a short time before, or part of a thousand-year-old dead redwood tree, or part of the snow cover on the mountain peaks. They may have just flushed the waste products out of the body of a deer or been formed by respiration in the body of a squirrel. Other molecules may have been many miles away in the Pacific Ocean, in a sewage plant, in a swimming pool, in a cactus on a desert. All reached the top of the mountain through the sun's energy. Their combined force as they tumble over the cliff in response to the pull of gravity comes directly from solar energy.

Courtesy National Park Service. Photo by Ralph H. Anderson directly back to the sun through the process of photosynthesis.

Water power comes from waterfalls, but the water got up on the hills and mountaintops when the sun's heat evaporated water and initiated the water cycle. Wind turns windmills, but air motion comes from differences in air temperature and, again, we're back to the sun.

Energy and matter are the two sides of the coin of life. Changes in matter always involve energy, and energy always moves through Earth's building blocks. Candy, sugar, chlorophyll, petroleum, peat, water, and air are all composed of elements.

As scientists learned more and more about the structure of the atom and the relationships between Earth and the sun, their understanding of energy increased. Finally, in the early part of the twentieth century, Albert Einstein, a Nobel-prizewinning German physicist who later would leave Nazi Germany and become a United States citizen, wrote his famous equation for energy: $E = MC^2$.

E in this equation stands for energy; M stands for mass, or the quantity of matter; and C is the speed of light, which is approximately 186,000 miles per second. Since the speed of light never varies, Einstein's equation, which is part of the theory of relativity, says that the amount of energy released is directly related to the amount of matter involved in the action. For example, when two molecules of sugar are oxidized, they produce twice as much energy as one molecule of sugar when it is oxidized.

Generally, $E = MC^2$ is applied to atomic energy, and implies that every part of the mass of an element is being used to form new elements in the atomic process. When the nucleus of an element like radium or uranium is split to form new elements like lead and radon, we call the process atomic fission or nuclear fission. When the nuclei of two atoms of hydrogen (H) are joined to form one atom of helium (He), we call the process atomic fusion or nuclear fusion. The energy from changes in atoms is called atomic energy or nuclear energy.

Some atomic fission goes on very slowly deep in Earth's interior, and is responsible for the great heat that melts and reshapes rocks and causes volcanoes, hot springs, and geysers.

Atomic fusion takes place on the sun. In fact, it is now believed to be the only source of the sun's energy. At first it was thought that the joining of two hydrogen atoms to form an atom of helium went on only at the sun's core, but recent photographs, which have been taken through powerful telescopes, and reports that have come back from space probes indicate that this kind of atomic activity also goes on at the sun's surface. As more and more is learned about the sun, we realize how fortunate we are to be 93 million miles away from it—near enough to receive life-supporting amounts of energy, but far enough away to be safe from the full blast of its power.

Whenever the structure of an atom of an element is changed, not only are tremendous amounts of energy released in the form of electromagnetic rays, some of which produce heat and light, but tiny particles of the atoms also break loose and move away from the area. Many of these particles are extremely damaging and destructive to all forms of life.

Energy constantly streams out from the sun in all directions. As the planets revolve around the sun, they pass through and are bombarded by these waves of energy.

As energy passes through space, it covers more and more area and becomes less and less concentrated. Many

The energy of the sun is so great that even when it is eclipsed it can cause eye damage or even blindness if one looks directly at it. Here a young man watches an eclipse by focusing the wide end of his field glasses on the sun and projecting its double image on a piece of white paper.

Courtesy Harry Wilks

more infrared rays (heat waves) strike a square foot of Mercury than strike a square foot of Earth; and, of course, the number striking Earth is much greater than the number striking a square foot of Pluto, whose orbit is approximately 3 billion miles farther away from the sun!

Even though the sun's energy is greatly dispersed by the time it reaches Earth, the biosphere could not exist if the ozone in the upper layers of the atmosphere did not screen out most of the ultraviolet rays, as well as the X rays and gamma rays produced by the sun's atomic activity.

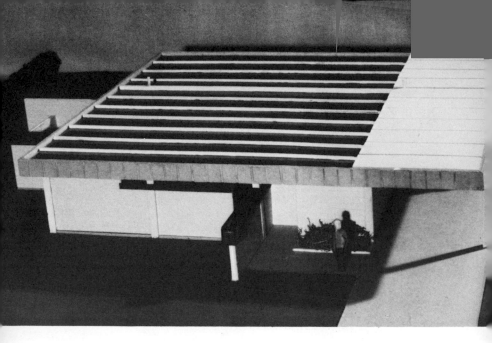

In this model of a house planned by Harold Hay of California, an eight-inch-deep pool of water on the roof is used to trap solar energy and provide both heat in winter and cooling in summer.

Courtesy Environmental Planning Consultants,
San Luis Obispo, California

A molecule of ozone consists of three atoms of oxygen. Ozone (O_3) is a very active substance that combines easily with other materials. In the lower levels of the atmosphere, it is irritating to lungs, eyes, and other delicate tissues. There it is a "natural resource out of place," but up in the stratosphere it is a guardian for all the things that make up Earth's biosphere.

Some of the infrared rays heat up some of the ozone in the stratosphere, and those rays never reach Earth; but most of the infrared rays, along with light waves, pass through the atmosphere to the lithosphere and hydrosphere.

Infrared rays and light waves must strike matter to become heat and visible light. Particles of matter are much farther apart in the gaseous atmosphere than they are in the solid lithosphere or liquid hydrosphere, so the atmosphere does not heat as well.

When the infrared rays strike a molecule of matter, they speed up the activity of the electrons. As electrons bump electrons in other molecules, they make these other electrons move faster, too, but at the same time they begin to slow down. With each transfer of energy, some energy is lost.

About 50 percent of the light that strikes a green leaf is absorbed by the chlorophyll of the leaf, but the sugar molecules produced by the process of photosynthesis probably contain less than 5 percent of the original energy. When these sugar molecules are oxidized through respiration, the energy is released in the form of heat. That heat may be converted to mechanical energy, or used to initiate chemical change, or used to keep your body at 98.6° F. (37° C.), but it will soon be dispersed and completely used.

You can watch energy "run down" if you spin a top. When you do this, you transfer to the top some of the energy that was released by oxidizing food in your body. The top spins vigorously at first, then slows down, and finally stops. Without a new input of energy, it will lie on the ground indefinitely.

Without a constant input of energy, you would quickly "run down" too. You can survive only minutes without respiration. Without constant input of sunlight, plants run down also. Once they use all their stored food, they die in a closed or a too-dark room.

All life would quickly grind to a halt if Earth's energy

When we talk about mechanical energy we are talking about two things: the energy of motion (kinetic energy) and the energy of position (potential energy). The little girl is swinging high because someone pushed her (kinetic energy). She will swing back because of her position in relationship to the ground (potential energy). Unless more energy is put in, eventually the swing will stop moving, just as a spinning top "runs down" or a basketball falls to the court or a muscle stops working unless there is a continued input of energy.

Courtesy Malcolm Frouman, Chelsea Clinton News

were not replenished daily by the sun. Eight minutes and twenty seconds after the sun stopped giving off energy, Earth would be a dark, rapidly cooling sphere. Soon water would freeze; winds would stop blowing; all liquids would solidify; and gases would turn to liquids and finally to solids. Only an occasional volcano would break the stillness and darkness as it spewed forth molten materials with a sudden burst of energy transferred upward to the top layers of the lithosphere from some deep, subterranean atomic reaction. This energy in the form of heat, light, and mechanical energy would soon be dissipated by the cold, dark, debris-covered lithosphere, where even the agents of decay would be unable to survive and carry on their important cleanup and recycling activities. Fortunately, the sun is not going to burn out in the immediate future.

Although energy is absolutely essential to every one of Earth's recycling activities, it is never recycled. Part of it may be temporarily stored in food, petroleum, or other substances, but once it is used, it is gone and must be replaced with more energy from the sun.

Ecologists say there are two interrelated basic principles on which the use of Earth's resources must be built if the biosphere is to continue to exist and flourish:

1. Energy moves in straight lines, and once it is fully used, it is gone.
2. Matter, which is composed of Earth's building blocks, moves in circular paths and is never destroyed.

Because energy does die out, it sets up some very important circular patterns called convection currents.

When infrared rays strike the lithosphere and hydrosphere, they are converted into heat energy, and the surfaces of water, rocks, soil, sidewalks, buildings, animals, and plants are warmed. They do not all become equally heated, however, for there is a vast difference in the efficiency of different substances in converting infrared rays to heat. In other words, the electrons of the molecules of some substances are much more responsive to heat waves than others, although all do respond to some degree.

Heat is transferred from these surface molecules to molecules below the surface and to molecules of the air adjacent to them. That this is the main source of the air's heat becomes evident when temperature readings are taken near Earth's surface, then twenty, a hundred, a thousand, or more feet above the surface. If the infrared rays were being converted to heat by air molecules, the air molecules nearest the sun would be hottest. Instead, the warmest air temperatures are nearest Earth's surface.

Whenever a gas is heated, the molecules move farther apart, with the result that a given volume of hot air contains fewer molecules than an equal volume of cold air. Since the weight of any substance at any spot in the universe is directly related to the kind and number of molecules it contains, a cubic foot of warm air weighs less than a cubic foot of cold air. In fact, it weighs so much less that it will begin to float upward. As the warm air moves up, cooler, heavier air sinks downward and takes its place.

In 1783 the first humans left the lithosphere and sailed through the atmosphere in a hot-air balloon invented by the Montgolfier brothers of France. Balloons of this type stayed afloat only as long as the air inside remained warm. As it cooled and the balloon contracted and become heavier than an equal volume of air, the balloon settled to the ground. This painting by Hugh Barclay Rich and Ethel Rich depicts the hundreds of people who turned out to watch the event.

Courtesy The Smithsonian Institution

Many Amish farmers in Lebanon and Lancaster Counties in Pennsylvania harness the sun's energy by using windmills that take advantage of the convection currents set in motion by unequally heated air.

The heated molecules lose energy as they move upward, and as their temperature drops, they move closer and closer together, until eventually they begin to join the downward-moving airstream.

When the heated air moves upward, it also carries water vapor upward. The cooling of the air as it is removed from its heat source results in the condensation of water and the formation of clouds. So the water cycle not only depends on the sun's heat; it operates because this energy, like all energy, is lost as it is used.

The same kind of convection currents arise at the equator, where the direct rays of the sun bombard Earth with more infrared rays than are received at places north and south of it, where the sun's rays strike at a slant. So heavy, cold air is constantly sinking and moving toward the equator, while light, warm air moves away from it. This movement of air, coupled with Earth's rotation, results in wind patterns.

The air movement also means that oxygen-rich air arising over tropical forests is distributed all over the world and makes life possible in cities, where the oxygen used by the people far exceeds the oxygen that plants in the area can produce. It also helps to make life possible on farmlands, on grasslands, and in forests in the world's temperate zones during the winter months, when green plants are inactive.

Other oxygen- and water-carrying convection currents move from the oceans to the land. Water molecules convert infrared rays to heat more slowly than many solid substances, but they lose heat energy more slowly, too. As a result, air molecules that are over the ocean in the daytime are heated less than air molecules contacting land, and air (wind) moves from the cooler ocean to the land. Since much of the world's oxygen is released by microscopic plants (called phytoplankton) that grow in the ocean all year, and by special chemical and physical changes that take place in some water molecules on the oceans' surface and somehow cause these molecules to split into their building blocks, oxygen and hydrogen, these ocean breezes help replenish the oxygen supply of both cities and the frozen winter countryside.

Water moves in similar convection currents. In the ocean, cold water flows from the poles toward the equator, and warm water flows over the cold and moves away from the equator. These moving masses of water influence the temperature and the moisture of the land they touch.

Warm water moving from the South American tropics swings across the Atlantic Ocean and warms the air off the islands and coast of northwestern Europe, while cold air coming from the arctic region chills the coast of Canada and New England. So although Goose Bay in Newfoundland and Shannon, Ireland, are the same distance from the equator, the former is the home of sled dogs and snowshoes and the latter has palm trees. Boston, Massachusetts, and Rome, Italy, are also on the same latitude, but again their climates are vastly different, due to ocean currents and patterns of air movement set up by the behavior of molecules of matter in response to the sun's energy.

In the Pacific Ocean, cold water from the Antarctic moves northward in the Humboldt Current. Air contacting this cold ocean current picks up very little water vapor. Later, as this cold wind moves over the hot, subtropical lands of Peru and northern Chile, it picks up moisture from the soil and creates a great coastal desert.

On the positive side of the ledger, this ocean current, like the cold current off Newfoundland, is an excellent place for phytoplankton to grow. Since phytoplankton are the food for many microscopic animals (zooplankton) which are eaten by larger animals, these areas are two of the best fishing zones in the whole world.

In places where water freezes, another factor enters the story of convection currents. We have already discovered that water is unique in its ability to dissolve almost all of the substances on Earth as well as in its ability to exist as a solid, a liquid, and a gas at Earth's temperatures.

In addition, water is unique in expanding when it freezes. In all other substances the molecules move closer and closer together as the substance is chilled and farther and farther apart as the substance is heated. In other words, other substances contract when cooled and expand when heated.

Water does this, too. As it is heated, it expands. As it is

Most of the precipitation in western Peru is in the form of snow on high mountains, where air moving upward is finally so chilled that it can no longer hold the water vapor it picked up as it moved over the land.

Because water expands between 4° and 0° Centigrade and therefore is less dense than warmer water (weighs less than an equal volume of warmer water), ice always forms on top of lakes, streams, and any other body of water.

During the winter many living things, like this three-pound lake trout caught by fishermen who cut a hole in Squam Lake, New Hampshire, remain active under the ice.

Courtesy State of New Hampshire. Photo by Dick Smith

cooled, it contracts *until* it reaches 4° C. (39.2° F.). Then
the pattern is reversed, and it begins to expand again until
it freezes at O° C. (32° F.). Because of this peculiarity,
in the fall, when the temperature of the surface water
drops to 4° C., the surface water is heavier than the water
below it, and it begins to move downward. As the lighter,
warmer water comes to the surface, it, too, is chilled to
4° C. Eventually, the whole pond or lake is cooled to 4° C.
Then the top layer cools still more, but this water, which
is below 4° C., is now less dense than the water below it,
and it stays on the surface. Soon ice forms, and the body
of water is still.

In the spring, when the ice melts and the surface water
changes from 0° C. to 4° C., it contracts and begins to
sink. This sinking water pushes water that is deeper in
the lake upward, and for several days or weeks these convection currents circulate the water and carry minerals
up from the bottom of the lake and dissolved oxygen
down from the surface of the lake.

Finally, as the surface water heats above 5° C., the
circulation from the bottom to the top and from the top to
the bottom stops until winter approaches and the water
temperature again drops and the top layer reaches that
heavy 4° C. temperature.

All over the world, air and water movements are set up
in patterns similar to these. They may be huge movements,
like a tropical storm covering great distances at great
speeds, or they may be small, like a local breeze blowing
from farmland or forest toward the hot asphalt, cement,
and bricks of the city; but regardless of their size, they are
all related to the sun's energy and its effect on Earth's
building blocks.

CHAPTER SEVEN

Food Chains

Matter and energy move from one organism to another in the biosphere in a cycle that is called a food chain. When green plants trap the sun's energy and build organic compounds from carbon dioxide and water, they are laying the foundation for all other life.

Animals feeding on the green plants use most of the organic compounds for energy, but a small amount of their daily intake is used to make other organic compounds for growth or for repair or replacement of tissues.

Most food chains have limitless variations. We can ima-

Unless there are beehives nearby or an abundance of wild bees, some fruit growers pay beekeepers to bring their hives to their orchards when the trees are in bloom; others keep their own hives and harvest both honey and fruit.

*Courtesy The New York State College of Agriculture and
Life Sciences, Cornell University*

gine one food chain if we start with an apple tree whose green leaves are trapping the sun's energy and producing food in the form of sugar from carbon dioxide and water.

Caterpillars feed on the leaves and digest the food that the tree has produced. Most of this digested food is oxidized. Some is used to build new caterpillar tissues.

A robin scoops up a mouthful of caterpillars and feeds them to a hungry baby. In the baby robin's body the caterpillar tissues are digested into amino acids and other small molecules. The amino acids are used to build protein for the bird's tissues, but much of the food from the caterpillars' bodies is oxidized for energy.

When the imaginary young robin is ready to fly, it sits on the edge of the nest, and it flutters to the ground—right beside a hungry cat! Even though both its parents give shrill directions and try to defend it, the bird cannot get away.

If the imaginary cat has had an abundance of protein recently, it doesn't need any more amino acids. Instead, when it digests the robin, it discards the nitrogen compounds in its urine. The carbohydrates are oxidized, producing carbon dioxide and water. The water evaporates and falls as rain, and the carbon dioxide becomes part of the atmosphere. The nitrogen compounds in the urine and any undigested parts that make up the feces are acted on by nitrogen-fixing bacteria and other biogeochemical recyclers.

This half-grown caterpillar has fed on young leaves every day, moving upward as new leaves unfold. Each feeding spot is a record of its appetite, which increases as it grows. Compare the size of its meals on lower leaves to its present feeding.

Thus the raw materials for building plant food are returned to the ecosystem, and the theoretical food chain is complete.

Green plants are the base for the food chain. Ecologists call them the producers. The rest of the biosphere is divided into two groups: consumers and decomposers. Some consumers, such as the caterpillars, eat plants and are called primary consumers. Others, like robins, eat the animals that eat plants. They are secondary consumers. Still others, like the cat, are tertiary consumers and they eat animals that eat animals that eat plants. Some animals might represent all three types of consumers. A robin, for instance, may eat a cherry, a caterpillar, and a spider.

Decomposers, the biogeochemical recyclers, act on every level of the food chain. If the leaves or other plant parts aren't eaten by the consumers, they are eventually broken down by the decomposers. Sooner or later all consumers die and are recycled by decomposers, as are all their body wastes while they are alive.

Sometimes this complicated pattern is called a web instead of a chain. Certainly, it is often an extremely tangled chain.

The importance of green plants in this cycle is obvious. They alone trap the sun's energy and start the chain of organic compounds by enabling the carbon in carbon dioxide to combine with hydrogen and other elements to form the substances of life. Since each consumer in the food chain oxidizes much of what it eats, the amount of photosynthesis required to keep the biosphere operating is tremendous.

In a day's time, a caterpillar can eat its own weight in leaves, and a one-week-old robin can eat its own weight in caterpillars. But neither doubles in weight. On an aver-

The fungus plants that produced these mushrooms obtained their food from green plants that had died and were buried in the ground. They are one of the decomposers that recycle materials which green plants need to grow.

In the meantime these mushrooms have been discovered by special mushroom flies, which are laying their eggs in the mushroom caps. In a few hours white, legless larvae will hatch from the eggs and start feeding on the mushroom tissues—decomposing a decomposer!

age it takes about ten grams of leaves to increase the caterpillar's weight one gram. It takes ten grams of caterpillar to increase a bird's weight one gram. This means that a baby bird is dependent on one hundred grams of leaves for a growth of one gram.

Mentally compare the weight of the food you consume in a year to your growth. For someone who enjoys statis-

While this brown thrasher is putting insects in one baby's mouth, two other babies are yelling for food. Parent birds of all kinds are constantly faced with open mouths and complaining babies who seem to have bottomless stomachs.

tics and playing with numbers, it might be interesting to calculate how many tons of green plants formed the base for one year's food intake.

In the mid-twentieth century, chemists developed techniques for making organic compounds that do not exist naturally on planet Earth. They took substances such as methane gas (CH_4) and set up conditions and energy transfers in such a way that some of the hydrogen atoms broke loose from the carbon atom and were replaced by chlorine atoms. These new substances are called chlorinated hydrocarbons.

About two hundred different chlorinated hydrocarbons have been developed. All of them are foreign to the ecosphere, and since they do not fit there, they kill living things. There are no recyclers that break them down. They are called persistent poisons.

The best-known chlorinated hydrocarbon is DDT. This is an abbreviation for dichloro-diphenyl-trichloro-ethane.

At the end of World War II, DDT was put on the market as an insecticide. At the time some scientists and conservationists felt that this was a mistake. They warned that it was a nonspecific cumulative poison. A nonspecific poison will kill many living things, instead of killing only certain specific things, and a cumulative poison is one that can be stored—it accumulates—in the liver and fatty tissues of animals' bodies.

But other people pointed to the fact that when DDT had been used as a powder to dust soldiers and civilians in Italy during World War II, it had stopped a typhus epidemic by killing the carriers—human lice and rat fleas. It had also been used on the *Anopheles* mosquito all over the world and had brought malaria under control. Obviously, they said, DDT could be used to make the world a pest-free, delightful place to live.

Admittedly, it could be a little dangerous, because in large enough doses it could kill any animal, but laboratory tests had shown that the small amounts needed to kill insects would not hurt mice, birds, or other large animals. So if people used it carefully, following the recommended dosage for insects, the benefits would far outweigh the risks.

Thus DDT went on the market. For a while, all looked rosy. Enthusiasm ran high. Homemakers sprayed roaches. Towns sprayed for mosquitoes. Foresters sprayed for gypsy moths. Dairymen sprayed cows to get rid of flies.

And then some disturbing things began to happen. Birds began to die. Some English farmers found hundreds of dead birds in one field. Some kinds of birds stopped reproducing or laid eggs with soft shells. Fish in ponds that had been sprayed for mosquitoes died. DDT was found in cow's milk and chicken eggs. The nonspecific cumulative poison was in the food chain. Even in the human one!

When DDT or any of the other chlorinated hydrocarbons is introduced into the food chain, it moves from organism to organism without changing its chemistry or its poisonous properties. Caterpillars that eat leaves sprayed with DDT die. If a parent robin picks up the dead or dying caterpillars and feeds them to its young in sufficient quantities, the young will die. Blowfly larvae feeding on the dead birds will consume the DDT and be poisoned by it.

Bacteria are not affected by the DDT, but neither do they affect it. So the soil bacteria break down the dead blowfly larvae into water, carbon dioxide, sulfur compounds, nitrogen compounds, and other small molecules and leave the DDT untouched. Rain running over the soil picks up the DDT and carries it. Eventually, most of it ends up in fresh water and oceans, where it is taken up by plankton and travels through many other food chains.

Early laboratory tests did not reveal the full effects of food chains on DDT concentration partly because they ignored the fact that DDT is a cumulative poison. As the laboratory tests showed, a bird eating an insect that died of DDT will not have a reaction to that amount of poison. What the tests did not show was that the bird will store the DDT in its body. As it goes on feeding on insects containing DDT in their bodies, it gradually accumulates more and more poison until finally the dose is lethal.

This cumulative action was not anticipated in laboratory tests, nor was the manner in which DDT and other chlorinated hydrocarbons would be carried to every niche in Earth's ecosphere. But when it was discovered that penguins in Antarctica had DDT in their liver and fatty tissues, the meaning of persistent poison began to be more fully understood, as did the carrying power of the great air and water currents set up by the transferral of energy.

Now many persons were ready to join the early protesters, though there were still people who laughed off the whole thing as unimportant. Why should they care about birds as long as they were free of mosquitoes and house flies?

But the small doses of DDT that once killed flies and mosquitoes didn't seem to affect these insects anymore. Big doses weren't killing them, either. Resistant strains of insects had developed. Even the scientists who had recognized the dangers of persistent nonspecific poisons had not anticipated that.

One house fly can lay six hundred eggs in her lifetime. It takes about twenty-four days for a fly to develop from egg to adult. Thus in warm climates fifteen generations of house flies can be born in one year. The potential descendants of one pair of house flies in fifteen generations would be 38×10^{36} if all survived and reproduced. Of course, the limits of food supplies, storms, parasites, predators, and other natural checks and balances keep the fly population down. Nonetheless the ability of house flies and mosquitoes to reproduce rapidly means that there is a much greater chance for genetic differences to be passed on to their offspring than there is in the much longer thirty-year generation of humans with much smaller families.

So when DDT spraying began, there may have been a few flies with immunity to the spray, or perhaps there were a few who reacted to a small dose of DDT as people

It is hard to imagine insects so small that they can live between the top and bottom layers of leaves, but you can often see these tiny larvae if you hold a leaf with a mine in it up to the light. (Most mines are made by tiny caterpillars, but a few are made by fly larvae or beetle larvae.)

The worst enemies of leaf miners are equally small or even smaller chalcid wasps which alight on the top of the leaf, drill a hole through the mine, and place an egg in the body of the miner. The wasp grub that hatches from the egg feeds on the miner and eventually kills it. Then it pupates in the mine and emerges as an adult to mate and lay eggs in other leaf miners.

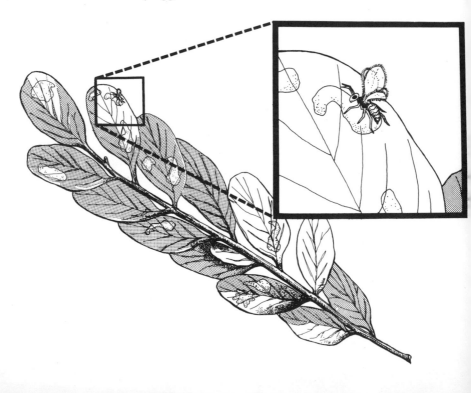

do to a vaccine. Either way, a few flies survived the spray program and lived to reproduce and pass on their resistance to their offspring, while the DDT-sensitive ones died. Soon only flies that could resist DDT remained to reproduce and pass on their characteristics to their offspring.

Mosquito populations reacted in the same way, so that some parts of the world, such as Guatemala and India, as well as some African and South American countries, experienced an increase in mosquitoes and malaria after a period of apparently successful spraying.

When resistant strains do develop and take over an area, they are frequently a much greater problem than the insects were before the spray program began, because the nonspecific DDT has killed many of the natural checks and balances.

Shortly after DDT was first applied to forest lands of the northeastern United States, the birch leaf miners increased in number. Year after year, every leaf of the white birch trees would have all the chlorophyll eaten from between the top and botton epidermis by these small insects. A tree that loses all its chlorophyll can live on stored food for a year or two, but if this happens for three successive years, the tree dies of starvation. In other areas black locusts and oaks were suffering in the same way.

DDT did not touch leaf miners, which were beautifully protected within the leaf, but it did kill the tiny chalcid wasps that fly about and lay their eggs in the bodies of leaf miners. DDT did not kill the many kinds of beetle larvae that feed under tree bark, but it killed the ichneumon and braconid wasps that parasitize them.

Today many persons are so far removed from the natural world and have heard so many pesticide advertise-

This twig has been split open to expose the larva of the flat-headed apple-tree borer, another insect that is well protected from sprays and therefore increases in numbers when the wasps that parasitize it are eliminated.

Courtesy The New York State College of Agriculture and Life Sciences, Cornell University

ments that they feel all invertebrate animals as well as many vertebrates are something that should be eliminated.

Without insects we would have no song birds, for even seed-eating birds feed their young an insect diet. Without insects there would be no trout and very few, if any, freshwater fish. Without insects the breakdown of dead plants and animals and the recycling of Earth's building blocks would be materially delayed. Without insects there would be no fruit crops and few vegetable crops.

Often we only learn about these relationships through

catastrophes. In 1970, an insecticide that was sprayed on cornfields in Washington and Oregon wiped out all the bees. In order to have a crop of apples and other orchard fruits, the states had to conduct an airlift of hundreds of thousands of honeybees from California the following spring.

Whenever you eat fruit, you owe a debt to a bee or some other insect as well as to a tree. You also owe a debt to a plant such as goldenrod or asters, for if bees did not store pollen from fall-blooming flowers in their hives, they would not have food for newly hatched larvae at the end of winter; and it is only these young bees, produced in the hive before the fruit trees bloom, that gather the early spring nectar and pollen and carry out the pollination that results in a fruit crop.

When prickly-pear cactus was introduced into Australia, it took over great areas. In the United States a caterpillar feeds in the fleshy stems and prevents the cactus from running rampant. Nobody appreciated that caterpillar until Australians discovered what a problem the cactus could be without its natural controls.

When the San Jose scale entered California on nursery stock from Pacific Islands, it almost wiped out the citrus industry. Entomologists who studied the scale in its native areas found that there it was not a problem, for black ladybird beetles overturned the scales and ate the insects "on the half shell." When the ladybird beetles were brought to California, they took over the control job, and the citrus industry was saved.

Sometimes the story doesn't end that happily. It is not always easy to repair damage. In 1967 it was discovered that Pacific Ocean coral reefs were disintegrating. Since coral reefs are large and complicated ecosystems involving many plants, invertebrate animals, and fish, this was a serious situation.

Studies of what was happening revealed that the crown-of-thorns starfish were reproducing at an alarming rate and feeding on the coral polyps. These big starfish had always been a part of the reef ecosystem in small numbers. What had caused their population explosion?

The same ecosystem had a marine snail that ate starfish eggs. The snail had a lovely shell, and for several years prior to the breakdown of the reef, shell collectors had been harvesting the snail and selling its shell. Who would have thought that the health of a coral reef might be related to a snail that ate starfish eggs? Or that many fish, a potential source of protein for hungry persons, could be lost because of snail shells in a curio shop?

Often the food chain is an essential part of the very complicated interrelationships of an ecosystem, and removing one link may upset the whole pattern.

The stony platform that millions of coral animals build under their bodies forms hiding places and habitats for a countless variety of other animals and plants.

In this reconstruction of a small portion of a coral reef from the Caribbean Sea, four kinds of animals besides different kinds of coral are visible: a sea anemone in left center, clusters of chimney sponges below it at lower left, a spiny lobster with only its long antennae extending from its grotto formed by three different kinds of coral on the lower right, and finally the Nassau grouper, half concealed as it waits for smaller fish to swim by and provide it with dinner. This represents only a small part of the variety of living things in the total reef.

Coral reefs everywhere are similar in their interrelated patterns and their importance as parts of the biosphere.

Courtesy American Museum of Natural History

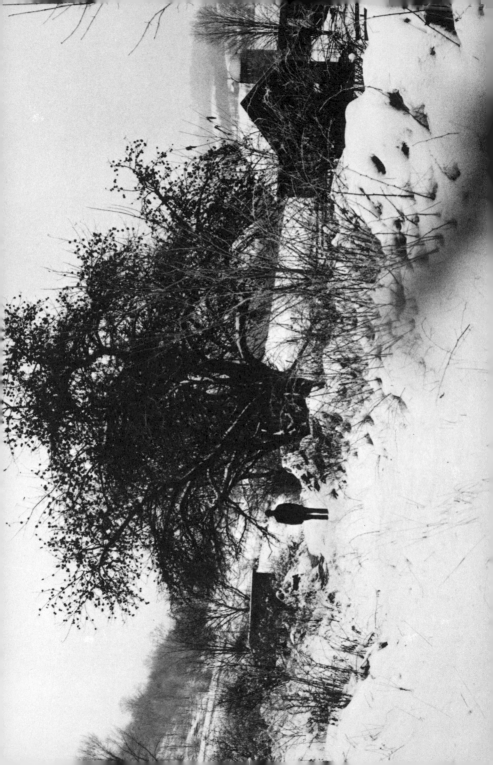

CHAPTER EIGHT

People and the Ecosystem

Whenever living things use matter and energy in carrying out their life activities, they change the environment.

Some of these changes are temporary. Heat escaping from the bodies of animals warms the air around them, but it soon disappears as the energy is transferred from one air molecule to another. Water vapor escaping from the tissues of living things temporarily raises the humid-

The deer have harvested the apples on the lower branches of this wild apple tree, a few other fur-bearing animals and insects have fed on others, but hundreds of apples still remain. Each apple can produce ten seeds. Think what would happen to this farm if every seed from this tree germinated.

Along the road are weeds and grasses with enough seeds to take over the land. Birds, mice, squirrels, and other animals will feed on them all winter and prevent these population explosions.

ity in that part of the ecosystem until it moves upward as part of the water cycle.

These changes can be experienced and recognized by all of us without using any scientific instruments. For example, notice the rise in temperature and the "stuffiness" that develops in a room full of people. When a room is full of people, the relative amounts of oxygen, water vapor, and carbon dioxide in the air also change. This, too, is a temporary and local change, which is offset by plant activities and air circulation.

In the same way all of the biogeochemical cycles create temporary changes. The length of time any of these environmental changes lasts is dependent on many factors, but the end result of these changes has been a series of checks and balances that keeps Earth's ecosphere healthy.

For hundreds of thousands of years, people fit into the pattern of natural recycling. When the human population was small and scattered, and people lived off the land by gathering plants and hunting, humankind affected the environment in exactly the same way as any other animal in balance with its ecosystem.

People are special, however. They have a brain that enables them to observe and ask questions, unlike other animals. Squirrels, for instance, have a working relationship with nut trees that neither they nor the tree realizes. Squirrels store nuts for future use. Nuts are seeds. Seeds are carefully packaged baby plants. Each year thousands of nuts buried by squirrels all over the world grow into young trees.

A tree produces many more seeds than it needs. In fact, if all the seeds from any one tree grew, there would be a disastrous population explosion of trees; conversely, if none of the seeds grew, trees would die out. When squir-

rels eat nuts, they are helping to prevent a population explosion; when they bury nuts, they are putting the
ready-to-grow young trees in a favorable habitat. This
partnership is one of the numberless delicate interrelationships of the biosphere.

When some of the nuts germinate, the squirrel hunts
for other food or goes hungry. Long ago, when people
stored seeds that germinated, they, too, had to hunt for
other food or go hungry. But at the same time some persons were asking, "Why?"

To find the answer to "Why?" requires reasoning and
observation. To use the answer requires imagination. This
involves asking other questions: "How can I change
things?" "What would happen if . . . ?"

Answers to questions and observations like this could
lead to several discoveries and to new practices, such as
roasting seeds before storing them, or storing seeds in dry
places. At the same time, someone else might have had an
entirely different thought, for this is one of the wonderful
things about human brains. Humankind's special reasoning ability can open the doors to limitless experiences
and ideas—many more than any one person could follow
in a lifetime. While one person thinks of one possibility,
another follows a different line of thought.

So in the long ago past, while some persons developed
better ways to store seeds, some other persons might have
thought, "If seeds grow when we store them in the
ground, maybe we could put them in a place where
we want them to grow, and then we would not have to
hunt for that kind of plant every year." Thus farming may
have started.

Once people no longer had to obtain all their food by
hunting and gathering, there was some time left over for

other things. Again, humankind's observing, thinking, reasoning, questioning brain led people in many directions. They developed ways to protect themselves from unfavorable weather and climate. They developed better ways to carry things, easier ways to move things, and tools to simplify their work. They became artists and musicians. They learned to harness the energy of fire and domestic animals and finally of electricity and the atom.

All these changes could never have occurred if record keeping had not also been developed. Possibly the first record keeping was a blaze on a trail that showed, "This is the path we followed," or a pile of rocks that indicated, "This is a good place to return to; food was abundant." Paintings on walls, notches in sticks, knots in twine —all were record-keeping devices. Although they seem primitive to us today, each reinforced people's memory and eliminated trial and error as the only way of learning. From these devices writing, books, and finally computers developed.

Instead of general insecticides, scientists today are developing ways to eliminate or control specific pests under specific conditions. Here a workman in Yakima, Washington, is blowing an insecticide composed of pyrethrin and a silica aerogel into the sewer system. Pyrethrin is an insecticide made from a garden flower. It is non-toxic to other animals and quickly decomposes. The silica aerogel soaks up the wax on the insects' bodies, so that they quickly dry out and die. In fact, the silica aerogel alone will kill roaches and other insects, and it is sometimes put into the walls of new buildings to make them insect-proof. It is excellent for this purpose, but it would not be a good insecticide to use in a garden area because it would kill useful insects as well as harmful ones.

Courtesy Niagara Chemical Division

Because of record keeping, we start with the accumulated knowledge of the centuries. It is sometimes said, "Each generation stands on the shoulders of its predecessors." Today we do not have to learn how to domesticate animals, hybridize plants, make a wheel, or harness electricity. Nor do we have to rediscover any other piece of information that has already been learned and recorded, so we can turn our attention to new discoveries.

As a result of this buildup of knowledge, humankind has been able to mold the natural environment and eliminate many factors that create discomfort, but in so doing, we have often disrupted and sometimes destroyed Earth's recycling patterns. In fact, the very comfort and safety that we seek has, in recent years, all too often been destroyed by air and water pollution and by other major disturbances in the delicate relationships of Earth's ecosphere. Sometimes, when we look at all our ecological problems today, we say, "How did it happen?"

Many different persons are seeking and giving answers. Some are well known. In 1962, Rachel Carson's book *Silent Spring* brought the problems created by persistent pesticides into focus, and initiated and strengthened many programs to protect the environment both here and in Europe. Rachel Carson was a biologist who recognized the importance of respecting the interrelatedness of all the ecosphere. She was also an extremely competent writer and a careful scientist, who spent five years visiting places where persistent pesticides were being used, reading reports, interviewing persons with experience in insect control, and collecting all kinds of data for the book.

In 1970, *Since Silent Spring*, written by Frank Graham, was published. This book describes the effect of *Silent*

Spring on people's thinking and reports on the progress that has been made as well as the problems that still exist.

Concern about the population explosion was brought to many people's attention in 1969 by another book, *The Population Bomb,* by Paul Ehrlich. Dr. Ehrlich is an ecologist and a population biologist at Stanford University.

For the past two centuries the world population has been increasing at an alarming rate. Whenever natural checks and balances are removed from any group of living things, they reproduce at a rate that makes them destroy their own environment. For instance, when the snails that ate starfish eggs were removed from the ecosystem of the Pacific coral reef, the starfish increased in huge numbers, but eventually, as the coral polyps were gobbled up and the reef decayed, the starfish were also without food and shelter.

Whenever the natural checks and balances are removed from any population, that population flourishes for a while, then suffers disaster because of overcrowding and lack of food, until finally new checks in the form of disease and starvation take over.

This has occurred repeatedly in the natural world. In the early part of the twentieth century, when wildlife managers decided to protect the 4,000 mule deer in the Kaibab National Forest in Arizona by removing the wolves, coyotes, and mountain lions, the deer herds increased rapidly. In twenty years they numbered 100,000. By this time they were eating all the plants on the ground. They were also eating all the young trees and bushes in the woods. As these disappeared, the mule deer stood on their hind legs and tore leaves and branches from trees. Finally, thousands died from starvation and from disease-causing organisms. Normally these organisms would only

Small hearses like this one, which was especially constructed for funerals of children at the end of the nineteenth century, remind us that childhood diseases were once a tragic factor in controlling human population.

Courtesy Museum Village of Smith Clove

slow down a well-fed deer, but they were fatal for the starving animals.

Dead deer were everywhere. In addition, the forests were badly damaged. Plants no longer held the soil in place when rain fell. The ecosystem was upset. It will be many years before all the scars of this population explosion disappear completely, even though predators have been reintroduced and the deer herd numbers about 10,000.

All living things have the potential for reproducing at a rate that far exceeds the carrying capacity of Earth. This surplus reproduction forms the basis for the food chain. It is part of the biosphere's secret of success. Primary consumers help keep plant poulations in balance, but secondary consumers prevent the primary consumers from increasing to the level where they would wipe out plant populations. Tertiary consumers are a check on secondary consumers.

People, like all other organisms, have the potential for reproducing at a rate that will upset the ecosystem. As long as the natural checks and balances existed, this did not happen. But one by one we have removed the checks. Wolves, cave bears, saber-toothed tigers, and other predators were a threat when people lived in caves, but no big predators serve as a population check today.

It took longer to bring parasites under control. In any eighteenth- or nineteenth-century cemetery, there are many small tombstones, which mark the graves of children. Most of these children died from diseases like diphtheria, scarlet fever, and polio. When your grandparents or great-grandparents were young, some children died in almost every family. This is no longer true. Very few children get these diseases today, and those that do are usually cured with modern medicine.

As a result the human population is spiraling upward. *The Population Bomb* emphasizes the fact that if we remove the natural checks, we must provide our own checks—in the form of smaller families—if we do not want to suffer from worldwide food problems and the breakdown of Earth's ecosphere.

A third book, *The Closing Circle*, written in 1971 by Barry Commoner, a professor of biology at Washington University in St. Louis, discusses the damage done to the environment by the production of chemicals that do not recycle, by atomic fallout, and by other industrial processes. It also stresses the economic and social policies that cause environmental problems.

The theme of each of these books is different, yet all point to one basic fact. People have been so busy exploring, inventing, manipulating, and exploiting the various parts of the planet Earth that they have failed to realize that natural cycles with all their complicated interrelationships cannot be bypassed or ignored.

It is time to look at priorities and decide what is really important.

At the seventh annual Inter-Society Conference on Energy Conversion, held in San Diego in 1972, Charles Backus, an engineer from Arizona State University, reported on the feasibility of harnessing solar energy to produce electricity by covering roofs of homes with solar cells. He stated that enough energy can be trapped in this way in the southern parts of the United States to provide electricity for the whole country.

Doing this economically would require mass production of solar cells, which are expensive today, because they are only produced for scientific research. Money to mass-produce them is not available.

At almost the same time that this report was given, the United States Congress appropriated $500 million to start construction of a breeder reactor, which would produce atomic energy to generate electricity.

All atomic reactions produce radiation, and radiation breaks down extremely slowly and travels to every spot in the world, just as persistent poisons do, through food chains and in air and water currents. Just as proponents of pesticides claimed, in 1946, that the potential benefits of using chlorinated hydrocarbons far outweighed the risks, so electrical companies, most atomic scientists, and many politicians soft-pedal the risks of atomic energy while they extol the imagined benefits.

In the past thirty or forty years, a great deal of money has been invested in atomic research, and it is easier to continue this than to explore new ideas. In addition, people would feel better about the atomic bomb if the atom could be harnessed for the benefit of humankind.

Since many persons have begun to fight pollution, electric companies have been able to convince some of them that nuclear energy is much cleaner than fossil fuels. After all, pollution from the production of electricity using coal is visible. Radiation pollution is invisible. Only its results in genetic damage, an increase in cancer, and ultimate death of plants, people, and other animals can be observed.

Proponents of atomic energy say that most radioactive wastes will not get into the atmosphere. Rather, these wastes will be stored in heavy tanks, while scientists try to figure out what to do with them. Many tanks have already been disposed of in the oceans, with the hope that they will remain intact for enough eons for all radioactivity to have died out.

In 1971, Arthur R. Tamplin, an atomic scientist at the Lawrence Radiation Laboratory of Livermore, California, wrote an article that appeared in the *Atomic Scientist*. It caused the same kind of concern as Rachel Carson's *Silent Spring*. In the article he discussed the great risks that we are taking by the large-scale production of atomic energy, and he noted that all these risks are being run, not to replace petroleum as a source of electrical power, but to add to it, so we can double our electrical production in the next quarter of a century and thereby provide all citizens with more things to buy and energy to use. This is an alarming statement when we realize that in 1971 the United States, with 6 percent of the world's pop-

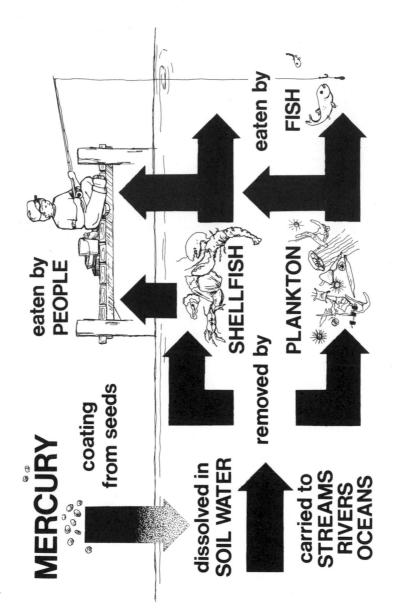

ulation, used 35 percent of the world's energy.

It is time for us to ask ourselves if we care to risk our health and the very existence of our descendants. Must we air-condition buildings so they are cooler in summer than the temperatures at which we heat them in winter, and heat them in winter so they are hotter than the temperatures at which we air-condition in summer? Should we produce aluminum cans to throw away, or wrap potatoes in aluminum foil when they actually bake better unwrapped, with oil on their skins? Should we pollute the environment producing electricity to convert our limited supplies of petroleum into plastics, to be used once and then tossed aside to contribute to the ton of rubbish that the average person throws away each year?

Twentieth-century scientists, "standing on the shoulders" of all their predecessors, have made discoveries that enable us to send people to the moon, to send delicate instruments to faraway spots in the solar system and beyond it, to tune in on electromagnetic waves and produce magic-like things such as television and radio telephones, which enable people to communicate with each other anywhere in the world and to send messages to space explorers with the speed of light.

The amount of poison required to kill a small organism will be taken into the body and stored in the tissues of a larger one without visible effect. However, as the larger organism continues to feed and store more of the poison, eventually it accumulates enough to kill it. Thus, as poisons move from organism to organism, we say that they undergo biological magnification. Persons on a daily diet of shellfish that have grown in water containing mercury compounds would develop reactions quickly, for they would be concentrating in their liver the accumulation of poison passed on from microorganisms to larger organisms in the food chain.

For a while, people were convinced that science and industry could solve all Earth's problems. Today we realize that those problems will only be solved when the health and well-being of the ecosphere, including all of Earth's people, become our primary concern.

Because the environmental problems of today's world are universal and polluters pollute more than their own dooryards, the United Nations Conference on the Human Environment was held in Stockholm, Sweden, in June 1972. It was the first worldwide conference concerned with planet Earth. It did not solve any problems, but just having a conference was a giant step forward. The conference focused the attention of people everywhere on the pollution problems of the world arising from the use of matter and energy without regard for their complicated interrelationships. It also focused attention on the equally complicated and interwoven cultural, economic, and social relationships of Earth's people.

Obviously, it is much easier to send people on trips to the moon than it is to solve the problems of the environment. Moon trips involve only a small number of persons who are tuned in on one specific set of problems. To solve environmental problems, scientists must deal with the hopes and aspirations of all the people of the world, with many kinds of ecosystems, and with every single use of matter and energy. Solving these problems is a tremendous task. It means that the knowledge of Earth's recycling must be applied to all of humankind's decisions. If this had been done in the past, recent catastrophes would have been avoided.

Before seeds were put on the market that had been coated with mercury to reduce loss caused by fungus, the following questions should have been asked and an-

The shellfish that breed in coastal wetlands like these are important not only as human food and as part of the food chain of the marine ecosystem, but they also are extremely important in retrieving valuable substances like compounds of potassium, calcium, and phosphorus before they are washed out into the inaccessible ocean depths.

Courtesy Chesapeake Bay Center for Environmental Studies

swered: "What will this killer do to essential soil bacteria?" "Where will this persistent poison move once it gets into water?" "What food chains might it enter?" Before mercury wastes from industry were discharged into Japanese waterways, similar questions should have been raised and answered.

Sometimes the relationships are not obvious. For example, phosphates are an important part of fertilizers. They stimulate plant growth. When they were first used in detergents, no one foresaw that the quantities that would ultimately get into rivers and lakes could eventually change the whole ecosystem, as they did in Lake Erie and many smaller waterways. By stimulating aquatic plant growth, the phosphates completely disrupted the physics and chemistry and therefore the biology of these parts of the hydrosphere.

Nor did anyone realize that the dredging and filling of coastal wetlands had already destroyed the breeding grounds of many shellfish, and had removed this important group of phosphate recyclers. As a result, large quantities of phosphate—which is an essential building block for all living things and exists only in limited deposits—are constantly being carried out to inaccessible ocean depths by rivers and streams.

Since Earth Day, April 22, 1970, more and more people have been demanding that environmentally damaging products and procedures be outlawed or changed. Individually, we can each influence the environment by our choice of products and the way in which we use energy and treat our small corner of the ecosphere, but we must band together in order to change legislation, to bring suit against offenders, to educate and inform other citizens, and to accept leadership responsibility. United action can make a difference.

Communities have changed their waste-disposal techniques, have dramatically reduced pollution levels, have replaced private transportation with good public transportation, and have made other changes that protect and improve the environment. These changes come only when people care enough to allocate tax money, to ask what is the best way, and to work together to build a human environment that meshes with and enhances the natural environment.

Some of the necessary changes also demand that people reexamine their values. For the past two or three decades, television has bombarded large segments of Earth's population with pictures of "the good life." The result has been a "gimme society" oriented to comfort and to material things (all wrapped in several layers of potential rubbish). Yet in spite of all our things, our labor-saving devices, our comforts, and our entertainment, more people than ever before find life without meaning.

Matter and energy cannot in themselves make life worthwhile, no matter how they are used. They can provide basic health, comfort, and safety, but beyond that, happiness and joy can only come from good human relationships and from a sense of accomplishment and dignity for all Earth's people.

Many people as well as countries might think seriously about the program that President Julius Nyerere of Tanzania has espoused. President Nyerere attended college in Uganda and in Edinburgh, Scotland. He knows the Western world and the African world. In 1961, he led Tanganyika to independence, and in 1964, he united Tanganyika and Zanzibar to form the country of Tanzania.

He is concerned with democratic government and with centralized economic planning that emphasizes equality for all citizens. To accomplish his goals, he has developed a program of villagization. Small farms and small, well-constructed homes keep their African characteristics while enabling each family to be comfortably housed and productive. President Nyerere regards big cities, with people without jobs, without contact with the natural world, and without dignity, as unhappy products of overindustrialization.

This does not mean that science and technology are bad in themselves. It does mean that industrialization, as well as agricultural practices that lose sight of human values and are carried out without respect for Earth's interwoven cycles, are destructive to all the biosphere and to humankind's spirit as well.

Epilogue—

Those Four-Letter Words

There are two ecologically destructive and devastating four-letter words which constantly creep into quiet discussions, angry arguments, scientific reports, and political speeches. These words refer to an almost limitless variety of situations, but in reality their meaning is always the same. Anywhere, anytime you can hear the words again and again:

"This used to be a good city until *they* moved in."

"Why don't *they* do something in the science department? We've run several studies in social science."

"That industry surely pollutes. Someone should make *them* clean up. *They* don't care what happens as long as *they* make money."

"Why doesn't the government step in? *They* are the only ones who can change things."

"This would be a good school if it weren't for *them*."

They and *them* used in this way are cop-out words that

permit us to shift responsibility onto someone else's shoulders.

No single person, or group of persons, is responsible for the conditions that exist today. We inherited Earth with its wonderful building blocks and recycling processes, with its beauty and its riches. We also inherited the things that humankind has done through the ages—the discoveries, the mistakes, the things that were successful, the things that backfired.

If we only look at the negative things and say, "*They* messed it up before I was born, and there's nothing I can do," then we are failing to take our responsibility because what happens right now is up to us. Furthermore, building a healthy ecosystem requires using knowledge and tuning into the economic social and cultural organizations that are also a part of our heritage.

In order to be a positive participating force at this time in history, we must be knowledgeable. That means we need to be able to distinguish between actual fact and propaganda. Many organizations or industries that want to keep the status quo have apparently jumped on the ecology bandwagon. Their ads may tell of their ecological contributions in glowing terms, or they may simply imply their contribution with beautiful colored illustrations. People who are poorly informed may be taken in by this kind of advertising, but the person who is informed will recognize it for what it is.

Books will provide good background material. Because things are constantly changing, periodicals are also essential sources of information. If your library does not have magazines that deal with ecology and conservation, ask your librarian to subscribe to some. Many of the best periodicals of this type are published by organizations that are carrying out programs to protect the environment. Some of these are:

American Nature Study Society
R. 1
Homer, N.Y., 13077

This organization was founded in 1908 to stress the importance of the natural environment at the time when the first major migration from farms to cities was taking place. Today it stresses the contributions, problems, and interrelationships of urban, rural, and wild environments. *Nature Study, A Journal of Environmental Education and Interpretation,* is published four times a year.

Audubon Society
1130 Fifth Avenue
New York, New York, 10028

The National Audubon Society was founded when concerned citizens banded together to fight the slaughter of birds for their feathers. After that victory was won, the emphasis gradually shifted to all of the interrelated forces that involve birds (which really is another way of saying "to the total ecosphere"). *Audubon Magazine* represents this total approach.

Massachusetts Audubon Society
Drumlin Farm
Lincoln, Massachusetts, 01773

Although this society's efforts are directed to the spearheading of educational programs and to providing leadership for many of the ecological battles of Massachusetts, the magazine *Man and Nature* is appropriate for any area of the Northeast, and its monthly newsletter contains some of the best articles to be found on issues that affect all people.

National Parks and Conservation Association
1701 Eighteenth Street NW
Washington, D.C., 20009

This association might be called the watchdog of the national parks and monuments. It provides leadership in fighting the many battles to keep the parks for all people and to prevent their destruction. Its *National Parks and Conservation Magazine* contains articles on the parks, their ecology, problems, history, uses, and beauty, as well as general articles on the environment.

National Wildlife Federation
1412 Sixteenth Street NW
Washington, D.C., 20036

Although the particular emphasis of this organization is on wildlife, it is well aware of the interrelated pattern of all life and tries to foster responsible thinking on the wise use of all our natural resources. It publishes *National Wildlife* as well as *International Wildlife,* a magazine that deals with the beauty and the problems to be found all over planet Earth.

The Sierra Club
1050 Mills Tower
San Francisco, California, 94104

This club, which started in the western United States, was one of the leaders in the battle against DDT and has become a very important national leader in environmental issues. It publishes *The Sierra Club Bulletin*.

Almost every state has monthly or quarterly publications directed to its voters. Some, like *The Conservationist*, published by the New York State Department of Environmental Conservation, Albany, 12201, are beautiful magazines with a wide range of interest beyond the state boundaries.

Others, like the four-page newsletter entitled *Air Pollution Notes*, published by the Cooperative Extension Service of The College of Agriculture and Environmental Science of Rutgers University, New Brunswick, New Jersey, 08903, may provide excellent news coverage on one specialized topic, as well as listing sources for additional information.

These publications are extremely helpful additions to any library whose readers are concerned with the environment. You can find out about this type of publication by contacting the extension service of your state university and your state Department of Conservation and the Environment.

Reading about ecology is not enough, however. Knowledge should lead to action. There are two ways to take environmental action. One is by our own life-style, the other by group effort. Both are important. Both need to be concerned with genuine problems and solutions and must not be satisfied with tokenism.

If we throw cans, bottles, candy wrappers, and hot-dog trays on the street and then complain about the filthy city; if we always use paper towels for mopping up spills, write on only one side of the paper, and never reuse paper products, then worry about our depleted forests; if we leave lights on and never do anything by hand that can be done electrically and then complain about the pollution from power companies; if we always travel in automobiles and never walk or bicycle or use public transportation, yet complain about air pollution—then we are hypocrites. We all must take responsibility for the environment in our daily living.

On the other hand, working alone on environmental betterment can be both frustrating and ineffectual. Many young people have discovered that they can accomplish a great deal by banding together.

In the fall of 1968, when students at the Thomas School at Rowayton, Connecticut, learned that they were living under a band of air which the Clean Air Task Force of the State of Connecticut had declared "constantly hazardous to the health level," they began to take an active interest in their environment. They discovered that the marshlands that they used as an outdoor laboratory with their biology teacher, Joy Lee, would soon be filled, and that all Connecticut marshes were doomed unless a bill which was in the legislature would be passed.

Since the Save the Marshlands Committee had been trying for ten years to have a bill of this kind passed, the outlook was not hopeful. However, Mrs. Lee and her students rolled up their sleeves and began to work.

They held a "Mourn-In" at the marsh to call the public's attention to the problem. They spent most of their out-of-school time working on the project, which included campaigning at the railroad station before school in the morning to obtain signatures on petitions and telegrams and enrolling the assistance of other students and other schools in the project. They attended the legislative hearings, and after the bill was passed they turned their

These young people are holding PYE buttons. The design is made up of a blue triangle symbolizing clean air and water over a green triangle representing open space, plant life, and natural resources, surrounded by a yellow edge denoting sunlight.

Courtesy Richard Harris, Manhattan Country School

attention to the problems that arise in putting a law into actual practice.

The name PYE, which stands for "Protect Your Environment," was selected by the students as their symbol. As other schools became involved, they also organized PYE clubs.

Today there are PYE clubs in grade schools, high

schools, and colleges all over the United States. There is no central headquarters or basic rules for these groups. PYE clubs are made up of young people who have banded together to bring about changes. In addition to PYE clubs, there are many environmental clubs operating under other names.

In Jersey City, a group of young people from Snyder

Cities do not save money when they charge people to use the dump. It costs more in terms of dollars and cents to clean up lots like this one, with the Statue of Liberty in the background, than the city makes at the dump. In addition, the cost in human values is unmeasurable.

High School operates a collecting station for glass, tin cans, aluminum cans, and newspapers in a parking lot in a shopping center every Saturday. Many persons who would not drive all over the city to deposit their "recyclables" in four different places are happy to drop off their flattened cans and bundled papers as they do their weekly shopping.

The money collected from this project is donated entirely to self-help projects in as wide a range as a project in Appalachia, a drug-rehabilitation program in the city, a farm-improvement program in Bangla Desh.

Other groups in other places have used money raised from similar projects to finance the travel, publication of materials, and visual aids required to educate people on some issue they were working on. Or the money might be used to enroll the school as a member of a local group that is fighting an environmental battle, or a national group like the Environmental Defense Fund.

The Environmental Defense Fund, 162 Old Town Road, East Setauket, New York, 11733, grew out of two persons' concern and imagination. Carol Yannocone, of Patchogue, Long Island, was disturbed when she read about fish that had died in Yaphauk Lake after it had been sprayed with DDT for mosquito control. Her husband, Victor Yannacone, was a lawyer. They decided to sue the Mosquito Control Commission of Suffolk County. Mr. Yannocone lined up scientists to provide the scientific facts for his case. This was the first of many court battles that finally led to the outlawing of DDT for general use.

At first the Environmental Defense Fund was made up of Victor Yannacone and a small number of scientists. Today it has a staff of lawyers and a large group of scientists. It also is composed of thousands of citizens who by

their membership not only help pay the high costs of the court battles but who have enabled the EDF to represent citizens in many environmental issues in all the courts of the land, including the Supreme Court.

Even if the fund had limitless financial backing, it could not bring these suits against major polluters if it could not say, "We represent this number of active concerned citizens who are backing us in this battle for a better environment."

Selecting the right approach to a problem is extremely important. Since Earth Day, thousands of vacant lots have been cleaned up, only to be refilled with the same kind of rubbish. In cities that charge homeowners and truckers to use the dump, as well as in cities where the sanitation department has a long waiting period before it will pick up any bulky object, vacant lots will be used as mini-dumps, and no amount of cleanup will change the basic problem.

Students who want to alter this kind of situation need to work on the passage of a new law and a change in policy. To succeed in this, they may need to enlist the help of other groups in the city.

Some cities have been extremely successful in creating clean-block associations. Here people in one city block band together to maintain their street. Many other side benefits often grow out of clean-block associations, such

When people band together, they can change things. When an old building was torn down in the Chelsea section of New York City, people in the nearby houses obtained permission to clear the lot and make a vest-pocket park for neighborhood use.

Courtesy Harry Wilks

This community park in Washington, D.C., was built when Walter Shapiro, a black college student, decided that young people of his neighborhood needed a place to play. After he obtained permission to use this vacant lot, the people of the community cleared it of weeds and debris and raised money for equipment and maintenance. One edge has garden space where children can have the pleasure of growing flowers and vegetables. Other sections are set aside for baseball and basketball courts.

Today this park is threatened because a realtor would like to use it for high-rise apartments. He says it is too valuable a piece of land to be used for a park.

Again we need to ask questions about values and the way we use land. What happens to Earth when too many people are packed close together? What happens to people? Is it good economy to save on parks and pay for crime, dope addiction, unhappy lives?

as vest-pocket parks, play areas, street fairs, educational scholarships, tree planting, windowbox contests. All of these are a part of the biggest gain of all: the feeling of community. As people join together to improve their environment, they become sensitive to each other as indi-

This symbol was designed for the United Nations Conference on the Human Environment held in Stockholm June 5–16, 1972. It is still in use today to remind us that there is only one Earth and that all of us are tied to it as well as to each other.

Courtesy United Nations

viduals. The people down the street are no longer "they" but have become partners in improving the environment.

Only as humankind joins together in many different kinds of activities, all designed to use energy, as well as Earth's building blocks, knowledgeably and responsibly, will we bring about the changes that are necessary to heal and protect the ecosphere so Earth's cycles can function successfully and all life can flourish on Earth, our home.

Index